# Digital Logic
# Techniques

**JOIN US ON THE INTERNET VIA WWW, GOPHER, FTP OR EMAIL:**

WWW:     http://www.thomson.com
GOPHER: gopher.thomson.com
FTP:     ftp.thomson.com
EMAIL:   findit@kiosk.thomson.com

A service of

# TUTORIAL GUIDES IN ELECTRONIC ENGINEERING

*Series editors*
Professor G.G. Bloodworth, *University of York*
Professor A.P. Dorey, *University of Lancaster*
Professor J.K. Fidler, *University of York*

This series is aimed at first- and second-year undergraduate courses. Each text is complete in itself, although linked with others in the series. Where possible, the trend towards a 'systems' approach is acknowledged, but classical fundamental areas of study have not been excluded. Worked examples feature prominently and indicate, where appropriate, a number of approaches to the same problem.

A format providing marginal notes has been adopted to allow the authors to include ideas and material to support the main text. These notes include references to standard mainstream texts and commentary on the applicability of solution methods, aimed particularly at covering points normally found difficult. Graded problems are provided at the end of each chapter, with answers at the end of the book.

Transistor Circuit Techniques: discrete and integrated (2nd edition) — G.J. Ritchie
Feedback Circuits and Op. Amps — D.H. Horrocks
Telecommunication Principles — J.J. O'Reilly
Digital Logic Techniques: principles and practice (3rd edition) — T.J. Stonham
Signals and Systems: models and behaviour — M.L. Meade and C.R. Dillon
Electromagnetism for Electronic Engineers — R.G. Carter
Power Electronics — D.A. Bradley
Semiconductor Devices: how they work — J.J. Sparkes
Electronic Components and Technology: engineering applications — S.J. Sangwine
Control Engineering — C. Bissell

# Digital Logic Techniques

## Principles and practice

## Third edition

**T.J. Stonham**
*Department of Electrical Engineering*
*Brunel University*

**CHAPMAN & HALL**
London · Weinheim · New York · Tokyo · Melbourne · Madras

**Published by Chapman & Hall, 2–6 Boundary Row, London SE1 8HN, UK**

Chapman & Hall, 2–6 Boundary Row, London SE1 8HN, UK

Chapman & Hall GmbH, Pappelallee 3, 69469 Weinheim, Germany

Chapman & Hall USA, 115 Fifth Avenue, New York, NY 10003, USA

Chapman & Hall Japan, ITP-Japan, Kyowa Building, 3F, 2-2-1 Hirakawacho, Chiyoda-ku, Tokyo 102, Japan

Chapman & Hall Australia, 102 Dodds Street, South Melbourne, Victoria 3205, Australia

Chapman & Hall India, Seshadri, 32 Second Main Road, CIT East, Madras 600 035, India

First edition 1984
Reprinted 1985, 1986
Second edition 1987
Reprinted 1988, 1989
Third edition 1996
Reprinted 1996

© 1984, 1987, 1996 T.J. Stonham

Typeset in The Republic of Ireland by Doyle Graphics
Printed in Great Britain by The Alden Press, Osney Mead, Oxford

ISBN 0 412 54970 0

A catalogue record for this book is available from the British Library

♾ Printed on permanent acid-free text paper, manufactured in accordance with ANSI/NISO Z39.48-1992 and ANSI/NISO Z39.48-1984 (Permanence of Paper).

# Contents

*this is a crap book*

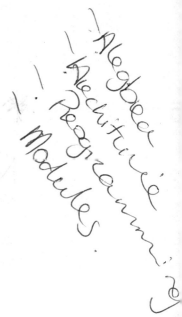

*Algebra*
*Architecture*
*Programming*
*Modules.*

# Preface to the third edition

In producing this third edition, attention has been given to the feedback from readers of the second edition and the result has been the inclusion of a completely new chapter on the design or arithmetic processors. The material has been produced in order to round off the range of topics needed by current students entering a course on digital electronics.

It is a fact that the range of material which the engineer needs to know increases rapidly with the passage of time and there is always pressure to cover more material in an undergraduate or Higher National programme. The necessity to retain all of the original basic material in the book has meant that the new topics have been chosen carefully in order not to make this introductory text too long. The other changes to the book have been in the form of increasing the number of worked examples in places where it was felt that the explanation was clarified by their presence. Additional exercises, together with answers, have also been provided.

# Preface to the first edition

The developments that have taken place in digital electronics over the past twenty years have no parallel in any other branch of engineering. A striking fact is that the real cost of hardware has halved every year whilst its complexity has, on average, quadrupled every three years. This is clearly illustrated by considering semiconductor memory. In 1964 one bit of storage would have comprised a single JK flip-flop constructed out of discrete components and cost the equivalent of twenty hours of a graduate engineer's time. Today single chips contain over one million bits of storage and have a cost equivalence of a tenth of a second of an engineer's time per bit. The increase in complexity and decrease in cost are both of the order of one million since the mid 1960s. Equivalent progress in, for example, the motor industry, would have provided us with luxury cars requiring only half a gallon of petrol for life having sufficient thrust to go into orbit and a price tag of about twenty pence.

With these developments as a backcloth, this introductory book aims to provide the necessary fundamentals of digital logic, whilst familiarising the student with design aspects and techniques at the system level. This is most relevant as current technology dictates that digital systems will be constructed from complex components, widely referred to as registers.

The starting point of the text is an objective comparison between analogue and digital representation of data. The student engineer must be aware that any given information processing task can be implemented either with analogue or digital electronics and in optimal design neither approach can be totally ignored. Digital methods of data representation are established in the opening sections. In line with the author's intentions to introduce the reader to contemporary aspects of the subject, the concepts of error detection and correction are included.

Boolean algebra provides a mathematical framework for digital electronics and has been used to link the propositional description of an operation to its implementation as a gate-level logic circuit. Combinational logic design from first principles provides the student with a vehicle to develop his understanding of digital electronics. Traditional methods of logic minimisation have been included as they are convenient 'pencil and paper' techniques. The emphasis placed upon them is, however, not high – reflecting their lessening importance in modern logic design. Cellular logic is presented as an alternative structure to canonical forms and is more relevant to VLSI implementation.

Sequential logic design is founded upon the flip-flop as an elementary building block. A strategy for the design of finite state machines based upon a generalised sequential logic structure comprising three data sets – the inputs, the internal states and the outputs – and two logic systems – the next state logic and the output logic – has been developed. Students can thereby create special-purpose sequential systems and also acquire a sound foundation for more advanced studies in Automata Theory.

The latter part of the book concerns the digital system as an entity. The implementation of logic systems in programmable hardware such as memories and arrays is examined and the specification of a system as a set of registers interconnecting via a bus structure is outlined. Fault diagnosis in a digital system has been defined and elucidated. This has often been regarded as a research topic but is now assuming increased importance in digital design. The author feels that these objectives must be introduced to the student engineer.

At the end of any engineering activity an artifact is produced. It is therefore essential that due attention is paid to the electrical properties of logic components. A comparison of the principal logic families is presented and the student's attention is drawn to potential problems which can arise when hardware systems are constructed.

When studying any subject, the student must experience for himself the fundamental concepts. To this end, numerous worked examples and marginal notes have been included in the text and graded problems with solutions given at the close of each chapter.

The author wishes to thank his Consultant Editors, Professor J.K. Fidler and Professor A.P. Dorey for their valuable advice throughout the preparation of this text. Thanks are also due to colleagues at Brunel University for counsel and encouragement. Finally, thank you to the Electrical Engineering students at Brunel for being a 'test-bed' for the approach to digital systems adopted in this book.

# Numerical representation of information

☐ To distinguish between analogue and digital quantities.
☐ To investigate the binary number system.
☐ To define and examine binary-coded decimal numbers.
☐ To specify the essential properties of a position sensing code.
☐ To introduce the concepts of error detection and correction in binary data.

In almost all activities, we are constantly dealing with quantities or measurements. This information is expressed in the form of numbers and can be processed in a digital system, provided it can be represented in an electronic form.

Information is almost always encoded in binary in a digital system and a wide range of codes exist. The choice of a particular code is influenced by the type of operation to be carried out on the data.

In this chapter, methods of representing numerical data in binary will be introduced. It is essential that the designer is familiar with fundamental coding techniques, as the form in which information is represented has significant influence on the design, performance and reliability of a digital system.

## Analogue and digital data

The first step in any data processing operation is to obtain information about the objects or phenomena of interest. The acquisition of information usually involves taking measurements on some property or characteristic of a system under investigation. In order to evaluate and assess the system, the measurements are monitored, scaled, compared, combined, or operated on in various ways. It is therefore essential that we have standard ways of representing our information.

Measurements can be divided into two broad categories. An analogue measurement is continuous and is a function of the parameter being measured. Conversely, a digital quantity is discrete and it can only change by fixed units.

Figure 1.1 shows two beakers being filled, one from a dripping tap and the other from a trickling tap. The build-up of water in the first beaker is typically digital, as the smallest change in volume is equal to one drip and the value increases in steps. In the second beaker, the volume of water increases continuously with time. It has an analogue property.

All drips are assumed to be the same size

Analogue and digital electronic circuits can be devised to process data. Analogue circuits such as operational amplifiers are cheap and powerful, but the analogue approach has two main disadvantages. The real world problem must have a precise electronic model and accurate measurements need to be made on that model.

For details on op-amps, refer to Horrocks, D.H. *Feedback circuits and Op-amps* (2nd edn, Chapman & Hall, 1990)

In digital electronic circuits, there is no need to make precise time-dependent measurements as a problem is modelled by a set of rules based on logic, which

Fig. 1.1  (a) A digital or discrete system. (b) An analogue or continuous system.

The resolution of a digital system determines the smallest detectable change in its parameters. It is dependent on the number of bits used to represent quantities.

will be developed throughout this book. One major drawback of digital circuits is resolution. If the resolution is increased by reducing the minimum step changes allowed in the parameters, the amount of electronic circuitry must increase.

The electronics engineer should be aware that analogue, digital or a hybrid combination of both methods may be relevant to a particular information processing operation. Neither approach should be precluded as any one method may provide a significantly simpler solution to a given problem.

Information is all about numbers. A number is a label with a unique meaning, and the remainder of this chapter will examine ways of representing information.

**Exercise 1.1**

Which of the following items involve analogue or digital quantities?
(1) Traffic flow
(2) Battery voltage
(3) Telephone exchange
(4) Temperature
(5) Vehicle speed
(6) Size of a pumpkin
(7) Pips in a pumpkin
(8) A number

## Number systems

### The decimal number system

A number system uses a set of symbols, known as digits. The total number of different symbols used by a given system is its base or radix. In the decimal number system, 10 symbols are used. They are 0, 1, 2, 3, 4, 5, 6, 7, 8 and 9 giving the base of the system as 10. The symbols are ordered and a change between two adjacent numbers is a unit or quantum. A complete number is the weighted sum of the powers of the base. Take, for example, the number 471. The base raised to power 0 is weighted by 1 (the right-hand symbol). The base to power 1 is weighted by 7 and so on. Hence the number is equal to

$$(4 \times 10^2) + (7 \times 10^1) + (1 \times 10^0)$$

The symbols themselves have a value and there is also a value with their positions. The complete number is the sum of the product of each digit value and its positional value. A general expression for the magnitude of a decimal number is

$$D = d_n 10^n + d_{n-1} 10^{n-1} + \ldots + d_1 10^1 + d_0 10^0 + d_{-1} 10^{-1} + \ldots + d_{-n} 10^{-n} \quad (1.1)$$

and the number is represented by the symbol string

$$d_n d_{n-1} \ldots \ldots d_1 d_0 . d_{-1} d_{-2} \ldots d_{-n}$$

where $d_i \in (0,1,2,3,4,5,6,7,8,9)$.

$\epsilon$ reads 'belong to'.

'Digit' is the Latin word for 'finger'.

The decimal number system no doubt evolved because humans have 10 fingers and in the early mechanical computers and calculating machines attempts were made at using base 10 numbers. Unfortunately there are very few devices that have 10 natural states, one for each decimal digit. In contrast, there are many systems where two discrete states can be detected without requiring precise measurement. Some typical binary state pairs are on/off, present/absent, high/low, true/false, positive/negative, stop/start, to quote a few.

Because of the ease with which two distinct states can be represented, the base 2 or binary number system has been universally adopted for electronic digital processing systems.

### The binary number system

In the binary number system, there are only two symbols, 0 and 1, and the base is therefore 2. The value of a pure binary number is given by the polynomial

$$B = b_n 2^n + b_{n-1} 2^{n-1} + \ldots + b_1 2^1 + b_0 2^0 + b_{-1} 2^{-1} + \ldots + b_{-n} 2^{-n} \quad (1.2)$$

and the number is written as the string

$$b_n b_{n-1} \ldots b_1 b_0 . b_{-1} b_{-2} \ldots b_{-n}$$

where $b_i \in (0,1)$ and is a binary digit or bit.

Numbers of the form shown in Equation 1.2 are called pure or natural binary numbers. Other binary codes exist that do not have this polynomial form.

$b_i \in (0,1)$ means $b_i$ is either 0 or 1.

### Conversion between decimal and binary numbers

A given quantity can be represented by a number in any base. The digit strings will differ, depending on the base being used, but the magnitude of the numbers will always be identical. Hence, given a number in base $a$, there is one and only one number in base $b$ that has the same magnitude.

The most important bases are 2, which is used in digital systems, and 10, which is used in everyday communications. A binary number can be converted into a decimal number by evaluating its polynomial, given in Equation 1.2.

Convert $11011_2$ into decimal.

*Solution*  The polynomial of the binary string is

$$(1 \times 2^4) + (1 \times 2^3) + (0 \times 2^2) + (1 \times 2^1) + (1 \times 2^0)$$

The decimal value of the polynomial is

$$16 + 8 + 0 + 2 + 1 = 27_{10}$$

Hence $11011_2 = 27_{10}$.

---

In order to convert from decimal to binary, a successive division procedure is used. In the general case, decimal number $D$ is identical to binary string

$$b_n b_{n-1} \ldots b_1 b_0$$

Hence

$$D = b_n 2^n + b_{n-1} 2^{n-1} + \ldots + b_1 2^1 + b_0 2^0 \tag{1.3}$$

Dividing each side of Equation 1.3 by the new base (which is 2) gives

$$\frac{D}{2} = b_n 2^{n-1} + b_{n-1} 2^{n-2} + \ldots + b_1 2^0 + b_0 2^{-1} \tag{1.4}$$

The left-hand side of Equation 1.4 consists of an integer part and a remainder where

$$\text{Integer} \left| \frac{D}{2} \right| = b_n 2^{n-1} + b_{n-1} 2^{n-2} + \ldots + b_1 2^0$$

and

$$\text{Remainder} \left| \frac{D}{2} \right| = b_0$$

The least significant bit of the binary string is therefore the remainder after the first division.

A further division of Integer $|D/2|$ by 2 will give $b_1$ and the process can be repeated until the complete binary string has been calculated.

In general, when converting from a smaller to a larger base use a polynomial expansion (see Worked Example 1.1). When converting from a larger to a smaller base use successive division (Worked Example 1.2).

---

Convert $19_{10}$ into binary.

*Solution*

$$D = 19$$

| Divide by 2 | $\dfrac{D}{2} = \dfrac{19}{2} = 9$ Rem 1 | $\therefore b_0 = 1$ | 2 is the base of the new number. |
|---|---|---|---|

Divide by 2 $\qquad \dfrac{D}{2} = \dfrac{19}{2} = 9$ Rem 1 $\qquad \therefore b_0 = 1$ 2 is the base of the new number.

2nd division $\qquad \dfrac{\text{Int } |D/2|}{2} = \dfrac{9}{2} = 4$ Rem 1 $\qquad \therefore b_1 = 1$ Rem means remainder.

3rd division $\qquad \dfrac{4}{2} = 2$ Rem 0 $\qquad \therefore b_2 = 0$

4th division $\qquad \dfrac{2}{2} = 1$ Rem 0 $\qquad \therefore b_3 = 0$

5th division $\qquad \dfrac{1}{2} = 0$ Rem 1 $\qquad \therefore b_4 = 1$

Hence $19_{10} = 10011_2$. Note the order of the bit string.

---

A binary number equation less than 1 can be converted to decimal by calculating its polynomial (Equation 1.2). A decimal number less than 1 can be converted to binary by a successive multiplication of its fractional part by the new base (2). The integer part of the resulting decimal number after each multiplication is a coefficient of the binary number, starting with the most significant bit. If a decimal number having both an integer and a fractional part is to be converted to binary, both parts must be processed separately. Use successive division for the integer part and successive multiplication for the fractional part.

In converting fractional parts, note that

$$0.1_2 = 0.5_{10}$$
$$0.01_2 = 0.25_{10}$$
$$0.001_2 = 0.125_{10} \quad \text{etc.}$$

---

Convert $0.8125_{10}$ into binary. **Worked Example 1.3**

*Solution*

Multiply by 2: $\qquad\qquad\qquad\qquad 0.8125 \times 2 = 1.625$

Multiply the fractional part by 2: $\quad 0.625 \times 2 = 1.25$

Multiply the fractional part by 2: $\quad\; 0.25 \times 2 = 0.5$

Multiply the fractional part by 2: $\qquad 0.5 \times 2 = 1.0$

The result is 0.1101. Note the order of the bit string.

---

Convert $0.1001_2$ to decimal. **Worked Example 1.4**

*Solution* The polynomial value of the binary string is

$$1 \times 0.5 + 0 \times 0.25 + 0 \times 0.125 + 1 \times 0.0625 = 0.5625$$

---

### Non-pure binary codes

*Binary-coded decimal*

A binary number system always uses only two different symbols, but the value of the numbers need not be based on a polynomial series. In an electronic system, the information is in binary form but when it is output to a user, decimal is more convenient.

A form of coding that enables a very simple conversion between binary and decimal is binary-coded decimal or BCD. In this case, each digit of a decimal number is replaced by 4 bits corresponding to its pure binary value.

---

**Worked Example 1.5**

Convert $45_{10}$ to BCD.

*Solution*

$$D = 45$$

Replace each digit with 4 bits of binary:

$$D = 0100/0101$$

Hence $45_{10} = 1000101_{BCD}$.

---

To convert from BCD to decimal, the binary word is divided into groups of 4 bits starting from the least significant bit and the decimal equivalent of each group obtained.

---

**Worked Example 1.6**

Convert $1010011_{BCD}$ to decimal.

*Solution*

Divide BCD into groups of 4:    0101/0011

Decimal value of each group        5/3

Hence $1010011_{BCD} = 53_{10}$.

---

BCD coding results in input and output devices being very simple as the value of each decimal digit is dependent on only 4 bits. The arithmetic process, however, becomes more complex as the rules of binary arithmetic no longer apply and certain pure binary numbers are not allowed. The above BCD coding is sometimes referred to as 8421 BCD, as this is the positional value of the bits within each group of 4.

If a code is simply referred to as BCD, assume that it is 8421 weighted, i.e. it uses pure binary values of the decimal digits.

Many other forms of BCD exist. A decimal digit need not necessarily be coded in pure binary, but it will always require 4 bits. A total of 16 different values ($2^4$) can be represented on 4 bits and 10 values are required to encode the set of decimal digits. In theory there are $16!/(16 - 10)!$ different ways of selecting 10 codewords from 16. This amounts to approximately $3 \times 10^{10}$ different BCD codes. The vast majority of these codes are quite random, and less than a hundred are weighted codes where each bit has a fixed weight. Some of the more common weighted codes are (4,2,2,1), (7,4,2,1) and (5,2,1,1). The weights need not be restricted to positive values and codes having some negative weights exist.

Some examples are $(7,4,-2,-1)$ $(8,4,-2,-1)$ $(-3,6,4,2)$. Code values for the decimal digits are shown in Table 1.1

The weighting indicates the value of the corresponding bit in the group of 4 bits representing the decimal digit.

**Table 1.1   Some weighted BCD codes**

The weights cannot *all* be negative.

| Decimal | 8421 | 4221 | 7421 | 5211 | 74−2−1 | 84−2−1 | −3642 |
|---------|------|------|------|------|--------|--------|-------|
| 0 | 0000 | 0000 | 0000 | 0000 | 0000 | 0000 | 0000 |
| 1 | 0001 | 0001 | 0001 | 0001 | 0111 | 0111 | 1010 |
| 2 | 0010 | 0010 | 0010 | 0011 | 0110 | 0110 | 0001 |
| 3 | 0011 | 0011 | 0011 | 0101 | 0101 | 0101 | 1100 |
| 4 | 0100 | 0110 | 0100 | 0111 | 0100 | 0100 | 0010 |
| 5 | 0101 | 0111 | 0101 | 1000 | 1010 | 1011 | 1101 |
| 6 | 0110 | 1010 | 0110 | 1001 | 1001 | 1010 | 0100 |
| 7 | 0111 | 1011 | 0111 | 1011 | 1000 | 1001 | 1110 |
| 8 | 1000 | 1110 | 1001 | 1101 | 1111 | 1000 | 0101 |
| 9 | 1001 | 1111 | 1010 | 1111 | 1110 | 1111 | 1111 |

---

Convert $739_{10}$ to $74-2-1$ BCD code.

**Worked Example 1.7**

*Solution*   Obtain $74-2-1$ codes for digits 7, 3 and 9 from Table 1.1:

$7_{10} \equiv 1000$

$3_{10} = 0101$

$9_{10} = 1110$

Hence $739_{10} \equiv 1000\ 0101\ 1110_{74-2-1}$.

---

If 1001 0001 0011 is an 8421 BCD word, what is the code for the same number in 5211 BCD?

**Worked Example 1.8**

*Solution*   Convert 8421 word to decimal:

$100100010011_{8421} \equiv 913_{10}$

Convert decimal number to 5211 code, with reference to Table 1.1:

$9_{10} \equiv 1111$

$1_{10} \equiv 0001$

$3_{10} \equiv 0101$

Therefore $913_{10} \equiv 1111\ 0001\ 0101_{5211}$.
    Hence $1001\ 0001\ 0011_{8421} \equiv 1111\ 0001\ 0101_{5211}$.

---

### Gray code

Quite often a digital circuit is used in a control system where the position of an object has to be measured. This can be done by encoding positions on the object and detecting the nearest position value. The edge of a conveyor belt, for example, could be encoded in binary using black and white markings as shown in Fig. 1.2.

Fig. 1.2   Pure binary position coding.

As the belt moves under a fixed transducer its position label can be read. If the positions are encoded in pure binary, errors can occur when the binary values representing adjacent positions differ by more than 1 bit.

Consider the belt moving from positions 1 to 2. Bit $b_0$ must change from 1 to 0 simultaneously with $b_1$ switching from 0 to 1. Electronic circuits can respond to changes of data with a microsecond and in practice $b_0$ and $b_1$ will not both change within this time. Bit $b_0$ will change either slightly before or after $b_1$ and erroneous position values 0 or 3 will be read when passing from positions 1 to 2. Similar errors will occur wherever adjacent positions differ by more than 1 bit, such as 3 to 4, 5 to 6, 7 to 8, and so on.

The problem can be overcome by using a reflected binary code, where adjacent values differ by only 1 bit. The most popular of these codes is the Gray code, which can be built up by progressively setting and reflecting the bits of the code word.

Starting from $0_{10}$ which is $0000_G$, the least significant bit is set to 1 giving

$$1_{10} = 0001_G$$

For $2_{10}$ and $3_{10}$, bit 1 is set to 1 and bit 0 reflected, giving

$$2_{10} = 0011_G \quad \text{and} \quad 3_{10} = 0010_G$$

For $4_{10} - 7_{10}$ bit 2 is set and bits 1 and 0 reflected and so on. The Gray code for decimal 0 to 16 is given in Table 1.2. It can be seen that adjacent values differ by only 1 bit. This makes the Gray code ideal for position sensing as the problems associated with the need for simultaneous changes of two or more bits never arises. A Gray position sensing code is shown in Fig. 1.3.

Switching is dependent on the alignment of the reading transducer, its linearity and the accuracy of the position code printing.

8

Table 1.2   The Gray code

| Decimal | Gray code | | |
|---------|-----------|---|---|
| 0 | 0 | | |
| 1 | 1 | | |
| 2 | 11 | Set bit 1. Reflect bit 0 | |
| 3 | 10 | Set bit 2. Reflect bits 1 and 0 | |
| 4 | 110 | | |
| 5 | 111 | | |
| 6 | 101 | | |
| 7 | 100 | | |
| 8 | 1100 | Set bit 3. Reflect bits 2, 1 and 0 | Imagine a mirror is placed under the reflected bits. |
| 9 | 1101 | | |
| 10 | 1111 | | |
| 11 | 1110 | | |
| 12 | 1010 | | |
| 13 | 1011 | | |
| 14 | 1001 | | |
| 15 | 1000 | | |
| 16 | 11000 | Set bit 4. Reflect bits 321 and 0 | |

Fig. 1.3   Gray position coding.

## Error detecting codes

When large amounts of data are being generated and processed in a digital system, there is always the possibility of some bits being corrupted. A one can be changed to a zero if part of an electronic circuit is not working, and errors can be introduced if data has to be transmitted over large distances by cable or radio.

### Parity codes

A simple method of error detection is to use an addditional bit, called the parity bit, with each binary word. Odd and even parity coding can be used. In odd parity coding, the parity bit is 0 if the total number of bits set to 1 within the word is

9

odd. If even, then the parity bit is set to 1 so as to make the total number, including the parity bit, odd.

The parity bit is generated before any data word is transmitted and a parity check can then be carried out in any subsequent part of the system. This involves a new calculation of the parity of the data word (excluding the parity bit). A comparison of the result with the parity received with the data word is then made and if they agree there is no single error in the data word.

Consider the binary word 010001. If it is to be transmitted with odd parity, the parity bit must be set to 1, giving

Now suppose bit $b_2$ is corrupted and becomes 1. The received word will be

A parity check on the data would now give a value 0, but the parity received is 1. This discrepancy can be detected and indicates that an error is present in the data. Unfortunately, with this simple parity scheme there is no way of telling where the error is. Furthermore, if two errors occurred within the same word, they would not be detected.

Even parity can also be used. The principle is the same as the odd parity scheme, except that the parity bit now makes the total number of bits set to 1 in the word even. Odd and even parity codes of pure binary numbers are given in Table 1.3.

**Table 1.3   Even and odd parity coded binary numbers**

| Binary | Even parity binary | Odd parity binary |
|--------|--------------------|--------------------|
| 0000 | 00000 | 10000 |
| 0001 | 10001 | 00001 |
| 0010 | 10010 | 00010 |
| 0011 | 00011 | 10011 |
| 0100 | 10100 | 00100 |
| 0101 | 00101 | 10101 |
| 0110 | 00110 | 10110 |
| 0111 | 10111 | 00111 |
| 1000 | 11000 | 01000 |
| 1001 | 01001 | 11001 |

*'m out of n' codes*

An alternative method of detecting errors is to use an '*m* out of *n*' code where *n* represents the total number of bits in a binary word, of which *m* must be set to

1. If more or less than $n$ bits are set to 1 then errors are present. The error detection circuitry has to count the number of bits set to 1 in a word and compare it with $m$. This is a relatively simple operation.

---

Devise a '2 in 5' code to represent the decimal digits 0 to 9. Each codeword must have 2 bits set and be 5 bits long. Valid codewords can be identified by counting in pure binary and using only those words that have 2 bits set.

**Worked Example 1.9**

| Count | Action |
|-------|--------|
| 00000 | Ignore |
| 00001 | Ignore |
| 00010 | Ignore |
| 00011 | Valid code equivalent to $0_{10}$ |
| 00100 | Ignore |
| 00101 | Valid code equivalent to $1_{10}$ |
| etc. | |

*Solution*   The full code is

| '2 in 5' | Decimal |
|----------|---------|
| 00011 | 0 |
| 00101 | 1 |
| 00110 | 2 |
| 01001 | 3 |
| 01010 | 4 |
| 01100 | 5 |
| 10001 | 6 |
| 10010 | 7 |
| 10100 | 8 |
| 11000 | 9 |

---

In the following bar code, a wide black bar represents $1_2$ and a narrow black bar $0_2$. If the bar code uses '2 in 5' coding, what is its decimal value?

**Worked Example 1.10**

*Solution*   Scanning the code from left to right, in groups of 5 bits, we obtain

00110, 10001, 01010, 11000, 00110

Referring to the '2 in 5' code table in Worked Example 1.9, we obtain the following decimal code:

$$26492_{10}$$

Additional information is encoded in the wide and narrow spaces in the bar code.

**Exercise 1.2**

Devise a '3 in 5' code for decimal digits.

**Error correcting codes**

Error detection signals can be used to stop any operation in a digital system and thereby prevent any false results or operations being generated. If, however, the errors can be detected and corrected, the processing operations can continue.

The principles of error detection can be extended to enable errors to be both detected and automatically corrected. Two elementary techniques will now be examined, which enable single errors to be corrected.

To detect and correct a single error within a data word, the number of parity bits must be increased. Consider a 4-bit binary word

$$b_3 b_2 b_1 b_0$$

Parity bits need to be generated for each group of 3 bits giving

$P_1$    parity for $b_3 b_2 b_1$
$P_2$    parity for $b_3 b_1 b_0$
$P_3$    parity for $b_2 b_1 b_0$

The total information which has to be transmitted is therefore

$$P_3 P_2 P_1 b_3 b_2 b_1 b_0$$

and in any subsequent error correction, all parity bits are recalculated and compared with the received values. Each bit in the received word can be corrupted and its binary value changed, giving a possibility of seven different single errors, which can occur in either the data or the parity sections. Suppose bit $b_0$ is corrupted. The parity check values of $P_2$ and $P_3$ will not then agree with the received values. A failure on parity bits $P_2$ and $P_3$ will only happen if an error occurs on $b_0$.

The data is binary and therefore there are only two possible values for each bit,

**Table 1.4 Error locations on a 4-bit data word with three parity bits**

| Parity failures | Error location |
|---|---|
| $P_2 P_3$ | $b_0$ |
| $P_1 P_2 P_3$ | $b_1$ |
| $P_1 P_3$ | $b_2$ |
| $P_1 P_2$ | $b_3$ |
| $P_3$ | $P_3$ |
| $P_2$ | $P_2$ |
| $P_1$ | $P_1$ |

so the error can be corrected by inverting the value of $b_0$. Table 1.4 shows the error correcting actions required on a 4-bit data word with three parity bits.

**Worked Example 1.11**

Detect and correct the error in the odd parity coded word 1110100. The format of the word is $P_3P_2P_1b_3b_2b_1b_0$ where $P_3$ is the parity of $b_2b_1b_0$, $P_2$ the parity for $b_3b_1b_0$ and $P_1$ the parity for $b_3b_2b_1$.

*Solution*   A parity check on the data section of the word gives

Use $P'_1$ for the parity check to distinguish it from $P_1$ the received parity.

$$P'_3 \quad b_2b_1b_0 = 0100$$
$$\therefore P'_3 = 0$$
$$P'_2 \quad b_3b_1b_0 = 1000$$
$$\therefore P'_2 = 1$$
$$P'_1 \quad b_3b_2b_1 = 0010$$
$$\therefore P'_1 = 0$$

Comparing parity check with parity received

$$P'_3P'_2P'_1 = 010 \quad \text{and} \quad P_3P_2P_1 = 111$$

Parity failures occur on $P_3$ and $P_1$.
    The error is therefore located at bit $b_2$.
    Hence by inverting the value of bit $b_2$, the correct codeword is obtained, which is

See Table 1.4.

    1110000

*Parity block checks*

An error correcting parity check on each data word can take a considerable amount of time. It also requires a significant increase in the number of bits needed to transmit a given amount of information. If the occurrence of errors is known to be infrequent, a parity block check system can be used, which will detect and correct one error in a block of data words.

    Each data word has a single parity check digit and a parity word is generated at the end of the block. Each bit in the parity word is a check on the corresponding bits in all the data words in the block. Consider a block of three words: 101, 100 and 010. Using even parity, the data block becomes

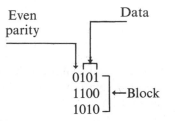

The parity word at the end of the block is obtained by calculating the parity of each column.

$$
\begin{array}{c|c}
0 & 101 \\
0 & 100 \\
1 & 010 \\
\hline
\text{Even parity word} \rightarrow 1 & 011
\end{array}
$$

If an error occurs in a data word, its own parity check will fail, together with one of the bits in the parity word. The error in the data is at the intersection of the row and column containing the failed parity checks.

---

**Worked Example 1.12**

The following block of data is received. It has been encoded with odd parity (MSB) and the last word is a parity block check. Identify and correct the error in the data. The data is 10000, 01111, 00100, 11100, 11010.

*Solution*   The data block as received, is

$$
\begin{array}{lll}
\text{Word 1} & 1 & 0000 \\
\text{Word 2} & 0 & 1111 \\
\text{Word 3} & 0 & 0100 \\
\text{Word 4} & 1 & 1100 \\
\hline
\text{Odd parity block} & 1 & 1010 \\
\text{check word} & \uparrow & \\
& \text{Odd parity bit} &
\end{array}
$$

Generating the parity from the data alone we get

$$
\begin{array}{r|l}
1 & 0000 \\
*1 & 1111 \\
0 & 0100 \\
1 & 1100 \\
\hline
0 & 1000 \\
& *
\end{array}
$$

Remember, the least significant bit is 'bit 0'.

The parity failures indicated by * can be identified and the intersection of the row and column locate the error. The error occurs in bit 1 of word 2 – it should be 0.
The corrected data block is therefore

The parity check data is discarded once the error has been located and corrected. Either use the received parity or recalculate parity from the corrected block.

$$
\begin{array}{r|l}
1 & 0000 \\
0 & 1101 \\
0 & 0100 \\
1 & 1100 \\
\hline
0 & 1010
\end{array}
$$

---

**Summary**

In this chapter, we have examined a wide range of binary codes that can be used to represent numbers. We, as humans, use the decimal number system, but a binary system is more suited to a machine implementation because two distinct states such as high and low voltage can readily be identified without having to make precise measurements. Methods of converting between binary and decimal number systems have been developed.

Conversion between decimal and pure binary can be quite laborious, as each bit may be dependent on all the decimal digits, and vice versa. BCD overcomes this problem as each decimal digit is determined by the value of a group of 4 bits. There are many BCD codes and weighted versions are generally used. The weights may have negative values. The most common BCD code is 8421 weighted, where each decimal digit is replaced by its value in 4-bit binary. BCD codes lead to simple input/output circuits but require complex arithmetic systems.

A class of code that facilitates the measurement of position must only change by 1 bit as the code is incremented or decremented. The most common position sensing code is the Gray code.

Finally, error detection and correction techniques were introduced. This is a vast and complex topic and the subject of much current research. Two relatively simple schemes for error detection were considered. The parity method uses an extra bit per word and the 'm in n' code only allows m bits to be set to 1 in an n bit word. If the number of parity bits per word is increased, errors may be detected and corrected. Alternatively, a parity word at the end of a block of data enables errors to be located. The concept of error correction is very exciting to the digital engineer. It enables him or her to correct data errors within digital circuits. Errors can arise when information is transmitted over long distances. Interference is likely to be present and corrupt the data, but error correcting techniques enable the correct information to be regenerated. In the home, the digitally recorded compact audio discs use error correcting codes. When a scratch or dirt corrupts the signal, error correction can restore the original signal and thereby remove the annoying clicks that would otherwise appear on the final sound output.

## Problems

1.1 Convert the following to binary

    (i) $141_{10}$
    (ii) $0.72265625_{10}$
    (iii) $21.8125_{10}$

1.2 Convert the following to decimal

    (i) $111001_2$
    (ii) $0.001011_2$
    (iii) $101.101011_2$

1.3 How many places are required to represent decimal numbers in the range 1–1000 in

    (i) binary
    (ii) base 8 (octal)
    (iii) base 16 (hexadecimal)
    (iv) $8421_{BCD}$?

1.4 Convert the following to base 8

    (i) $182_{10}$

(ii) $1011110101_2$

(iii) $47.75_{10}$

1.5 Convert the following to base 16. Use symbols A–F for digit values 10–15.

(i) $1011100111000111_2$

(ii) $4732_{10}$

(iii) $254.03125_{10}$

1.6 Convert the following numbers to 8421 BCD

(i) $4723_{10}$

(ii) $1010111011_2$

(iii) $AA4E_{16}$

1.7 Convert the following numbers to $74 - 2 - 1$ BCD

(i) $47_{10}$

(ii) $9F_{16}$

(iii) $101101_2$

(iv) $100001000111_{8421}$

1.8 Convert the following numbers to '2 in 5' code

(i) $36_{10}$

(ii) $101110_2$

(iii) $4A_{16}$

(iv) $10010111_{8421}$

1.9 Detect and correct the errors in the following data

| | | | | | | |
|---|---|---|---|---|---|---|
| | 1 | 0001 | | | 1 | 0110 |
| | 0 | 1011 | | | 0 | 1000 |
| (i) | 0 | 1100 | | (ii) | 1 | 1100 |
| | 0 | 0000 | | | 0 | 1111 |
| Even parity → | 1 | 0010 | | Odd parity → | 1 | 0011 |
| word | ↑ | | | word | ↑ | |
| | Even parity bit | | | | Odd parity bit | |

1.10 Devise a method whereby two numbers, coded in 8421 BCD, can be added together giving the answer in BCD.

1.11 Design an alternative position sensing code to that given in Fig. 1.3.

1.12 Explain the difference between weighted and non-weighted binary-coded decimal and give an example of each type of code.

# Operations on binary data          2

---

☐ To define combinational, sequential and storage operations.
☐ To relate logic to propositional statements.
☐ To define a truth table.
☐ To examine binary connectives and thereby define the fundamental logic gates.
☐ To apply Boolean algebra to logic design.

**Objectives**

---

We saw in Chapter 1 that information could be encoded into binary, and quantities represented by strings of symbols, where each symbol can only have one of two values. We shall now examine relationships between binary variables.

Electronic logic operations are governed by the same principles and laws as the propositional statements we use in everyday life, as the latter can be regarded as binary functions that are either true or false.

**Logical operations**

An electronic circuit that operates on binary data is called a digital logic system. The circuits within the system that carry out the elementary logical operations are called gates. Three general types of logic system can be identified, namely:

(i) combinational logic system,
(ii) sequential logic system,
(iii) storage system.

Any logic system can be represented by a black box, having a set of input lines that receive binary data, and one or more output lines, as shown in Fig. 2.1. If the system is combinational, it has the following property: the output data is a logical function of the value of the input data at the instant the output is measured. If the inputs are represented by the set $\mathbf{I}$ and the outputs by the set $\mathbf{Z}$, then

$$\mathbf{Z}_t = f(\mathbf{I}_t) \tag{2.1}$$

where $\mathbf{Z}_t$ is the output and $\mathbf{I}_t$ the input at time $t$. $f(\mathbf{I}_t)$ represents the function carried out by the logic system. Consequently a given input $\mathbf{I}_t$ will always generate

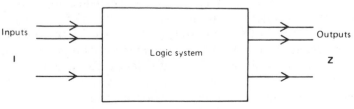

Fig. 2.1   A general logic system.

the same output from a combinational logic circuit, irrespective of when it is applied to the system.

An adder circuit is an example of a combinational system. If the inputs represent two numbers, say 2 and 3, we would expect an immediate output of 5, and furthermore, the output must be 5 each and every time the inputs are 2 and 3.

A given output can be produced by more than one set of inputs. Inputs 4 and 1 would also give output 5.

A sequential circuit can also be represented by Fig. 2.1 as it also has inputs and outputs. The function of a sequential circuit differs from a combinational logic system in one important respect. The outputs depend not only on the present inputs, but on all the previous inputs over a specific operating time.

Hence for a sequential system

$$\mathbf{Z}_t = f(\mathbf{I}_1 \mathbf{I}_2 \dots \mathbf{I}_t) \tag{2.2}$$

where $\mathbf{I}_i$ is the input data at time $i$.

As the output of the circuit depends on the values of the present and all previous inputs, a given input at time $t$ does not always produce the same output.

Sequential behaviour can be observed in an accumulator circuit that calculates the cumulative sum of all the inputs. Consider the sequence of numbers 1, 4, 2, 5, 2 being input, one at a time, into an accumulator. The outputs are shown in Table 2.1. Each output is the running total of the present and all previous inputs. At time 3, the input is 2 and the resultant output is 7, but at a later time (5) an identical input gives an output of 14. This could not happen in a combinational circuit and provides a means of distinguishing between them. If at any time, a logic system gives different outputs for the same input, the system must be sequential.

**Table 2.1   An accumulator operation**

| Time | Input | Output |
|------|-------|--------|
| 1 | 1 | 1 |
| 2 | 4 | 5 |
| 3 | 2 | 7 |
| 4 | 5 | 12 |
| 5 | 2 | 14 |

In the accumulator circuit, the output is dependent on the magnitude of the previous inputs. In other sequential systems, the behaviour may be dependent on both the magnitude and the order of the input sequence. A telephone handset is an example of the latter. Consider the case where five digits of a six-digit number have already been dialled. The sixth and final digit can have one of only 10 values, but the connection is made to one out of a possible maximum of $10^6$ subscribers, depending on the value and order of the previous five digits.

The third type of logic system is the storage operation. A storage device has two distinct functions. It can hold information, where data is input and stored but nothing is output, or alternatively information within the memory can be accessed, giving an output without requiring a data input. In memories where more than one number can be stored, additional input information is required to specify the exact location in the memory where the data is to be stored. This input is called the address. Another control input is also necessary to set the memory into its read or write mode. The terminals on a memory circuit are shown in Fig. 2.2.

'Read' means 'access data already stored in memory'.

Fig. 2.2 A memory system.

'Write' means 'store data in memory'.

The three types of logic systems are inter-related in a most interesting way. A combinational logic circuit can be converted into a sequential system by applying feedback from some of the outputs to form additional internal inputs. A simple sequential circuit has memory properties and finally a memory circuit can be organized as a combinational logic function. These applications will be examined in the following chapters.

The fundamental electronic circuits common to all types of logic systems are the gates. A logic function is implemented with one or more gates and the relationships between functions are governed by the laws of Boolean algebra.

## Logic and propositional statements

The formal analysis of binary systems was first investigated by the nineteenth century English mathematician, George Boole, long before the advent of electronics and computers. Boole's logical algebra was developed to test the validity of propositions and verbal statements, but his theorems are directly relevant to the operations which are performed on binary data in electronic systems.

Consider the simple statement about the weather: 'It will snow if the temperature is low and it is cloudy.'

The three variables are 'snow' (the output), 'low temperature' and 'cloudy' (the inputs). Each of these conditions can only be true or false (i.e. true if it is snowing or false if it isn't).

This weather system can be described by a Boolean equation, with the variables S for snow, L for low temperature and C for cloudy, giving

$$S = L \text{ AND } C \tag{2.3}$$

The logic function between the two input variables is AND, which can be identified in the verbal statement. The symbol for AND is . and equation 2.3 would normally be written

$$S = L . C \tag{2.4}$$

The symbol . may be omitted.
$S = LC$ is the same as $S = L . C$.

The equation states that S is true if, and only if, L is true AND C is true. It represents the proposition 'It will snow if the temperature is low and it is cloudy.'

**Truth tables**

Every Boolean equation has a truth table that lists the value of the output for each and every possible combination of inputs. If the output is a function of two variables there are $2^2$ or four possible input combinations. In general, for $n$ variables, the truth table will have $2^n$ input states.

The truth table for the propositional statement

$$S = L.C$$

can be obtained as follows. The four input values for L and C are (False, False), (False, True), (True, False) and finally (True, True).

The output S can be determined by substituting the input conditions into the statement. S can only be true if L is true and C is true. S is therefore false for all other input conditions and the complete truth table is given in Table 2.2(a). In electronic systems, two voltage levels would be used to represent truth and falsehood and they would be labelled with binary symbols 0 and 1. If truth is represented by logical 1 and falsehood by 0, the binary version of the truth table can be obtained (see Table 2.2b).

*The system is assumed to be binary.*

*The variables can only be true or false.*

**Table 2.2  Truth table for the AND function S = L . C**

| Output S | Inputs L | C | S | L | C |
|---|---|---|---|---|---|
| False | False | False | 0 | 0 | 0 |
| False | False | True | 0 | 0 | 1 |
| False | True | False | 0 | 1 | 0 |
| True | True | True | 1 | 1 | 1 |

(a) Truth values          (b) Binary symbols

**Worked Example 2.1**

Obtain a Boolean equation and its truth table, which specifies the suitability of an applicant for employment with a corporation, requiring employees to be either

  (i) unmarried females under 25 years, or
 (ii) married males under 25 years, or
(iii) over 25 years.

*Solution*  Let E represent the suitability of the applicant. The input variables are married status M, where M true is married and M false is unmarried, sex S, where S true can represent male and S false, female, and A for age where A true means the applicant is over 25 years.

$\overline{M}$ *reads NOT M. When* $\overline{M}$ *is true, M is false.*

Condition 1 is NOT  M  AND NOT  S  AND NOT  A
Condition 2 is M  AND  S  AND NOT  A
Condition 3 is A

Therefore suitability E is ($\overline{M}$ AND $\overline{S}$ AND $\overline{A}$) OR (M AND S AND $\overline{A}$) OR (A)

Replacing AND with the symbol . and OR with +, the Boolean equation is

$$E = (\overline{M}.\overline{S}.\overline{A}) + (M.S.\overline{A}) + A$$

The truth table will contain eight input conditions as there are three input variables. Condition 1 is satisfied by input reference 0, condition 2 by input 6 and condition 3 by inputs 1, 3, 5 and 7. Hence the truth table is

The input reference number is the decimal value of the inputs where F = 0 and T = 1.

| E | M | S | A | Input reference |
|---|---|---|---|---|
| T | F | F | F | 0 |
| T | F | F | T | 1 |
| F | F | T | F | 2 |
| T | F | T | T | 3 |
| F | T | F | F | 4 |
| T | T | F | T | 5 |
| T | T | T | F | 6 |
| T | T | T | T | 7 |

## Binary connectives

A binary connective is an elementary function of two input variables. Consider a simple binary system having single output Z and two inputs A and B.

There are four possible combinations of truth values that can occur at the two inputs. They are FF, FT, TF, and TT. As the four possible input states are applied in turn, they will give rise to a sequence of four outputs, each of which can be true or false. Thus a total of $2^4$ or 16 different output responses is possible. Each set is a binary connective, and represents a unique function between the input variables and the output. The full list of connectives is given in Table 2.3. It contains all possible functions of two input variables.

F is false. T is true.

Connectives 0 and 15 are independent of the inputs and always give a constant output, whereas 3 and 5 are both independent of one variable, and have outputs identical to A and B, respectively; 10 and 12 are also only dependent on one variable, the outputs taking the opposite value of the inputs. This is the INVERSE or NOT function. The two propositional functions AND and OR occur as connectives 1 and 7, whereas 8 and 14, the inverse of these functions, are called NAND and NOR.

NAND means NOT AND.

Connective 6 is the EXCLUSIVE OR function. It differs from the OR operation in one important respect. The OR function is true if both inputs are true, whereas the EXCLUSIVE OR would be false under these input conditions. Both functions are true when only one input is true.

OR is sometimes known as INCLUSIVE OR.

Table 2.3  **Binary connectives between inputs A.B and output Z**

| A | F | F | T | T | Function | Symbol |
|---|---|---|---|---|----------|--------|
| B | F | T | F | T |  |  |
| 0 | F | F | F | F | Universal falsehood | 0 |
| 1 | F | F | F | T | AND | A.B |
| 2 | F | F | T | F | — | — |
| 3 | F | F | T | T | Variable A | A |
| 4 | F | T | F | F | — | — |
| 5 | F | T | F | T | Variable B | B |
| 6 | F | T | T | F | EXCLUSIVE OR | $A + B$ |
| 7 | F | T | T | T | OR | $A + B$ |
| 8 | T | F | F | F | NOR | $\overline{A + B}$ |
| 9 | T | F | F | T | Equivalence | $\overline{A + B}$ |
| 10 | T | F | T | F | Not B | $\overline{B}$ |
| 11 | T | F | T | T | — | — |
| 12 | T | T | F | F | Not A | $\overline{A}$ |
| 13 | T | T | F | T | — | — |
| 14 | T | T | T | F | NAND | $\overline{A.B}$ |
| 15 | T | T | T | T | Universal truth | 1 |

The logic designer must be aware of the difference between OR and EX.OR. The distinction is not always obvious in natural language. Consider the following statements:

(i)  Safety limit for elevator is 12 persons <u>OR</u> 1 ton goods.
(ii) We will acept US dollars <u>OR</u> pounds sterling.

OR function can have any number of inputs. EX.OR is a function of only two inputs.

Statement (i) uses the EXCLUSIVE OR connective whereas statement (ii) is the OR function.

**Worked Example 2.2**

Which binary connective is being used in the following statement?

'For £10 I can buy one compact disc OR two cassette tapes.'

*Solution*  The statement was an OR connective. Assuming a CD costs £10 and a tape costs £5, the statement implies purchase of the CD and not the tapes or the tapes and not the CD. It is not possible to buy both the two cassette tapes and the CD. This is the EXCLUSIVE OR operation:

$$F = A \oplus B = A\overline{B} + \overline{A}B$$

## Logic gates

A logic gate is a two-state device that implements a binary connective. Most logic devices are electronic, although logic operations can be performed by mechanical, magnetic and even fluid systems. The fundamental gates are AND, OR and NOT and any logical statement can be expressed by using a combination of these three gates.

AND, OR and NOT form a universal set of gates from which all other logic functions can be constructed.

**Worked Example 2.3**

Express binary connective 13 in terms of AND, OR and NOT gates.

*Solution*   The truth table for binary connective 13 is

| A | B | Z |
|---|---|---|
| F | F | T |
| F | T | F |
| T | F | T |
| T | T | T |

The proposition is:

Output Z is true    if A is true and B is true
                      or A is true and B is false
                      or A is false and B is false

Expressing this statement as a Boolean equation, we obtain

$$Z = A.B + A.\overline{B} + \overline{A}.\overline{B}$$

which reads

Z equals A and B or A and not B or not A and not B

**Exercise 2.1**

Verify that all the binary connectives dependent on two input variables can be expressed with AND, OR and NOT gates.

Other gates include the EXCLUSIVE OR function

$$Z = A\overline{B} + \overline{A}B$$

and its inversion, which is the EQUIVALENCE function

This operation is sometimes called EXCLUSIVE NOR.

$$Z = AB + \overline{A}\,\overline{B}$$

NAND and NOR gates are the inversion of AND and OR, respectively. The principal logic gates, their symbols, equations and binary truth tables are summarized in Table 2.4.

23

**Table 2.4 Logic gates**

| Name | Symbol | Equation | Truth table A | B | Z |
|------|--------|----------|:---:|:---:|:---:|
| AND | | $Z = A.B$ | 0 | 0 | 0 |
| | | | 0 | 1 | 0 |
| | | | 1 | 0 | 0 |
| | | | 1 | 1 | 1 |
| OR | | $Z = A + B$ | 0 | 0 | 0 |
| | | | 0 | 1 | 1 |
| | | | 1 | 0 | 1 |
| | | | 1 | 1 | 1 |
| NOT | | $Z = \overline{A}$ | 0 | | 1 |
| | | | 1 | | 0 |
| NAND | | $Z = \overline{A.B}$ | 0 | 0 | 1 |
| | | | 0 | 1 | 1 |
| | | | 1 | 0 | 1 |
| | | | 1 | 1 | 0 |
| NOR | | $Z = \overline{A + B}$ | 0 | 0 | 1 |
| | | | 0 | 1 | 0 |
| | | | 1 | 0 | 0 |
| | | | 1 | 1 | 0 |
| EXCLUSIVE OR | | $Z = A \oplus B$ | 0 | 0 | 0 |
| | | | 0 | 1 | 1 |
| | | | 1 | 0 | 1 |
| | | | 1 | 1 | 0 |
| EQUIVALENCE (EXCLUSIVE NOR) | | $Z = \overline{A \oplus B}$ | 0 | 0 | 1 |
| | | | 0 | 1 | 0 |
| | | | 1 | 0 | 0 |
| | | | 1 | 1 | 1 |

## Boolean algebra

The principles of logic were developed by George Boole (1815–1884) who, along with Augustus De Morgan, formulated a basic set of rules that govern the relationship between the true–false statements of logic. Boole's work laid the foundation to what later became known as Boolean algebra. Nearly 100 years later, Claude Shannon, an American postgraduate at the Massachusetts Institute of Technology, realized that Boole's work was relevant in particular to the analysis of switching circuits in telephone exchanges and, more generally, formed a mathematical basis for the electronic processing of binary information.

The theory of electronic logic systems is often called 'switching theory'.

*Boolean theorems*

The theorems of Boolean algebra fall into three main groups:

(i) Logical operations on constants.

(ii) Logical operations on one variable.
(iii) Logical operations on two or more variables.

*Logical operations on constants*

Also known as Huntington's postulates.

| AND | OR | NOT |
|-----|-----|------|
| $0.0 = 0$ | $0 + 0 = 0$ | |
| $0.1 = 0$ | $0 + 1 = 1$ | $\bar{0} = 1$ |
| $1.0 = 0$ | $1 + 0 = 1$ | $\bar{1} = 0$ |
| $1.1 = 1$ | $1 + 1 = 1$ | |

In the AND operations, 0 dominates. If one or more inputs are at 0, then the output becomes 0.

In the OR expressions, a logical 1 on the inputs is dominant. The output becomes 1 if one or more inputs are at 1.

*Logical operations on one variable*

| AND | OR | NOT |
|-----|-----|------|
| $A.0 = 0$ | $A + 0 = A$ | |
| $A.1 = A$ | $A + 1 = 1$ | $\bar{\bar{A}} = A$ |
| $A.A = A$ | $A + A = A$ | |
| $A.\bar{A} = 0$ | $A + \bar{A} = 1$ | |

*Logical operations on two or more variables*

*Commutation*

$$A + B = B + A$$
$$A.B = B.A$$

The commutation rule states that there is no significance in the order of the variables. A OR B is identical to B OR A.

*Absorption*

$$A + A.B = A$$
$$A.(A + B) = A$$

The absorption rule can be used to simplify logic expressions.

*Association*

$$A + (B + C) = (A + B) + C = (A + C) + B = A + B + C$$
$$A.(B.C) = (A.B).C = (A.C).B = A.B.C$$

The association rule allows variables within the same logic operation to be grouped together in any order. For example, the OR of three variables A, B and C can be achieved either with a three-input OR gate, $A+B+C$, or by using two gates each having two inputs. The first gate can produce the OR of any two input variables and its output is then ORed with the third variable giving $(A + B) + C$ or $(A + C) + B$ or $A + (B + C)$. All four versions are logically identical.

$\overline{A + B}$ reads A NOR B.
$\overline{A.B}$ reads A NAND B.

$$\overline{A + B} = \overline{A}.\overline{B}$$
$$\overline{A.B} = \overline{A} + \overline{B}$$

De Morgan's theorems are most important in logic design. They relate AND to NOR and OR to NAND.

*Distributive laws*

$$A.(B + C) = A.B + A.C$$
$$A + (B.C) = (A + B).(A + C)$$

$(A+B)(A+\overline{B}) = A.A + A\overline{B}$
$\qquad\qquad\quad + AB + B\overline{B}$
$\qquad\quad = A + A\overline{B} + AB$
$\qquad\quad = A + A(\overline{B} + B)$
$\qquad\quad = A + A$
$\qquad\quad = A$

Distribution is a process similar to factorization in arithmetic.

*Minimization theorems*

$$A.B + A.\overline{B} = A$$
$$(A + B).(A + \overline{B}) = A$$
$$A + \overline{A}.B = A + B$$
$$A.(\overline{A} + B) = A.B$$

INVERSION is the same as NOT.

In Boolean algebra, as in arithmetic, there is an order of precedence of operations. Bracketed functions are evaluated first, followed by AND, then OR and finally INVERSIONS.

The expression $A + B.C$ means A ORed with the result of B AND C. It is *not* C ANDed with the output of A OR B. The latter would have to be written

$$(A + B).C$$

The digital systems engineer must be aware of the principles of Boolean algebra; however, it is not necessary to memorize all the theorems. The operations on logical constants and single variables are rather obvious and can easily be established from first principles and the associative and commutative properties occur in arithmetic. We shall also examine graphical and tabular methods of simplifying logic equations that replace the absorption and minimization theorems. De Morgan's theorems, however, are important in the design of logic systems. They do not have a parallel in arithmetic and must be learnt.

*Verification of Boolean theorems*

Boolean theorems can be verified by logical reasoning or by demonstrating equivalence relationships on a Venn diagram.

---

**Worked Example 2.4**

Show that $A + \overline{A} = 1$ using logical reasoning.

*Solution*

$$\text{Let} \quad F = A + \overline{A}$$

The equation reads

'iff' reads 'if and only if'.

$\qquad$ F is 1 iff A is 1 or not A is 1

But not A is 1 means A is 0.

$\qquad \therefore$ F is 1 iff A is 1 or A is 0

But A is binary and must be either 0 or 1, so one of the conditions for A is always satisfied.

$\therefore$ F is always 1
$\therefore A + \overline{A} = 1$

In a Venn diagram, the binary variables of a function are represented as overlapping areas in a universe. The OR function is the combination or union of areas and the AND function is the intersection or common part between two or more overlapping areas. The NOT function is the remainder of the universe outside a given area. Two functions are equivalent if they define identical areas on a Venn diagram.

Show that $AB + A\overline{B} = A$ by using Venn diagrams.

**Worked Example 2.5**

*Solution*

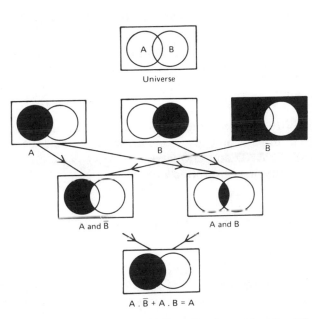

A . $\overline{B}$ + A . B = A

Final Venn diagram is identical to A.

A third method can be used to verify a Boolean identity. The truth tables for both sides of the equation are calculated, and if they are identical for all input values, then the identity is true.

Show that $AB + A\overline{B} = A$ by using truth tables.

**Worked Example 2.6**

*Solution*   Let $F = AB + A\overline{B}$. The truth table for this is

| A | F |
|---|---|
| 0 0 | 0 |
| 0 1 | 0 |
| 1 0 | 1 |
| 1 1 | 1 |

By comparing columns A and F, it can be seen that they are identical. Hence F = A, so

$$AB + A\overline{B} = A$$

Verify De Morgan's theorem $\overline{A.B} = \overline{A} + \overline{B}$ using truth tables.

*Solution*   Given $\overline{A.B} = \overline{A} + \overline{B}$. Let $F_1 = \overline{A.B}$ and $F_2 = \overline{A} + \overline{B}$.
The truth tables for $F_1$ and $F_2$ are

| A | B | A.B | $F_1 = \overline{A.B}$ |
|---|---|-----|------------------------|
| 0 | 0 | 0   | 1                      |
| 0 | 1 | 0   | 1                      |
| 1 | 0 | 0   | 1                      |
| 1 | 1 | 1   | 0                      |

and

| A | B | $\overline{A}$ | $\overline{B}$ | $F_2 = \overline{A} + \overline{B}$ |
|---|---|----------------|----------------|-------------------------------------|
| 0 | 0 | 1              | 1              | 1                                   |
| 0 | 1 | 1              | 0              | 1                                   |
| 1 | 0 | 0              | 1              | 1                                   |
| 1 | 1 | 0              | 0              | 0                                   |

Comparing $F_1$ and $F_2$ we see that they are identical for every input AB. Hence

$$\overline{A.B} = \overline{A} + \overline{B}$$

**The principle of duality**

A careful scrutiny of the Boolean theorems will reveal an interesting property. Most theorems occur in pairs and, given one theorem, the other can be obtained by interchanging all the AND and OR gates.

Take, for example, the first absorption rule

$$A + A.B = A$$

If the OR gate is replaced by AND and vice versa, the second absorption theorem is obtained:

$$A.(A + B) = A$$

This natural pairing is an example of the principle of duality, which states that for every logic system there exists a dual form that performs the same function. The dual form can be obtained by interchanging all the AND and OR gates and inverting every input and output.

Given the function F = (A + B).C, obtain its dual form and prove that the dual performs the same function as F.

*Solution*   Given F = (A + B).C.
The dual may be obtained by inverting all inputs and outputs and interchanging AND and OR gates. Hence

$$\overline{F}_D = (\overline{A}.\overline{B}) + \overline{C}$$

$F_D$ is the dual of F.

$F_D$ reads 'NOT A ANDed with NOT B and then NORed with NOT C'.

Inverting both sides to obtain $F_D$ gives

$$F_D = \overline{(\overline{A}.\overline{B}) + \overline{C}}$$

$F_D$ is of the form $\overline{X + Y}$ where $X = (\overline{A}.\overline{B})$ and $Y = \overline{C}$. But $\overline{X + Y} = \overline{X}.\overline{Y}$ by De Morgan's theorem. Hence

$$F_D = \overline{\overline{A}.\overline{B}}.C$$

Also

$$\overline{\overline{A}.\overline{B}} = A + B \text{ by De Morgan's theorem}$$
$$F_D = (A + B).C$$

Hence

$$F_D = F$$

## Truth table equivalence

Boolean algebra provides a mathematical foundation for binary information processing. It can be used to describe complex operations, prove identities and simplify logic systems. In practice, however, Boolean algebra is of limited value to the engineer. The success of a minimization or a proof of an identity depends largely on the person's expertise, and success cannot be guaranteed.

If we take two logic systems where one is considerably more complex than the other, it is important to know whether they perform the same process. If they do, then the engineer would always choose the simpler system. Boolean algebra can be used to establish identities; however, if two systems cannot be proved equivalent by algebra, the result is inconclusive. Either the systems are not equivalent or the engineer's algebra is inadequate.

A better, more reliable method of demonstrating equivalence between two systems is to generate their respective truth tables. If two or more systems have identical truth tables, then they perform the same function.

Truth table equivalence has already been used to verify Boolean theorems. It is, however, of great practical importance as it provides a quick and simple method of determining equality between complex logic systems.

If $F_1 = A(A + \overline{B}) + BC(\overline{A} + B) + \overline{B}(A \oplus C)$ and $F_2 = A + C$ determine whether $F_1 = F_2$ by means of truth tables.

**Worked Example 2.9**

*Solution* $F_1$ is a function of A, B and C. $F_2$ can be considered a function of A, B and C with its output independent of B.

Truth tables are

Remember $A \oplus C$ is the EX.OR function and equal to $A.\overline{C} + \overline{A}.C$.

| Inputs | | | Intermediate terms | | | Outputs | |
|---|---|---|---|---|---|---|---|
| A | B | C | $A(A + \overline{B})$ | $BC(\overline{A} + B)$ | $\overline{B}(A \oplus C)$ | $F_1$ | $F_2$ |
| 0 | 0 | 0 | 0 | 0 | 0 | 0 | 0 |
| 0 | 0 | 1 | 0 | 0 | 1 | 1 | 1 |
| 0 | 1 | 0 | 0 | 0 | 0 | 0 | 0 |
| 0 | 1 | 1 | 0 | 1 | 0 | 1 | 1 |
| 1 | 0 | 0 | 1 | 0 | 1 | 1 | 1 |
| 1 | 0 | 1 | 1 | 0 | 0 | 1 | 1 |
| 1 | 1 | 0 | 1 | 0 | 0 | 1 | 1 |
| 1 | 1 | 1 | 1 | 1 | 0 | 1 | 1 |

If the output cannot be calculated directly from the Boolean equations and the input values, evaluate any necessary intermediate terms first.

By comparing the output columns, we can see that $F_1$ and $F_2$ always have the same value for each and every output.

Hence $F_1 = F_2$.

**Exercise 2.2**

Use Boolean algebra to determine whether or not the following functions are equivalent:

$$F_1 = \overline{(A + B).(A + C)} + \overline{A + B + C}$$

and

$$F_2 = \overline{A}\,\overline{B}\,\overline{C} + ((A\overline{C}) \oplus \overline{B})$$

Confirm your finding by means of truth tables.

### Summary

Logic operations can be divided into three broad and inter-related groups. In a combinational operation, the output at a given time is a function of the inputs at that instant. In a sequential operation, the output is a function of the value and order of a series of inputs over a given period of time. A storage operation can hold input data and output it at a later time. Logic functions are similar to propositional statements and can be completely and uniquely specified by a truth table that gives the output values for each and every possible combination of inputs.

A binary connective is a function that relates the output of a binary system to its inputs. By considering all possible functions of two inputs the fundamental logic operations can be identified.

Finally, the relevance of Boolean algebra to logic design was assessed. Boole's theorems, which can be verified by logical reasoning, using Venn diagrams or truth table equivalence, are directly applicable to digital logic systems. However, in practice, truth table equivalence may provide a quicker and more efficient means of proving an identity than the use of Boolean algebra.

### Problems

2.1   Identify the following two-input functions

(i)   The output is 0 only when both inputs are 1.
(ii)  The output is 1 provided the inputs are different.
(iii) The output is 1 provided no more than 1 input variable is 1.
(iv)  The only time the output is 0 is when both inputs are 0.

2.2   Prove the following relationships using Venn diagrams

(i)  $A + \overline{A}B = A + B$
(ii) $\overline{A.B} = \overline{A} + \overline{B}$

2.3   Devise truth tables for

(i)  A three-input NAND gate.

(ii) a three-input NOR gate.

2.4 Devise a truth table for a logic system, with two inputs A and B, where the output F is equal to B if A is 0, and $\overline{B}$, if A is 1. What gate is required to implement this function?

2.5 Obtain truth tables for the following equations

    (i) $F = \overline{A} + B.\overline{C} + \overline{A.B}$

    (ii) $F = (A + B).(\overline{A} + \overline{B} + C)$

    (iii) $F = (A.B) + \overline{\overline{A} + \overline{B}}$

2.6 Prove the following identities using Boolean algebra

    (i) $\overline{(A + B).(\overline{\overline{A.B}})} = \overline{A}.\overline{B}$

    (ii) $(A + \overline{B}).(\overline{A} + \overline{B} + C) = A.C + \overline{B}$

    (iii) $(\overline{A \oplus B}) + (B \oplus C) + (A \oplus C) = 1$

2.7 Verify the identities in Problem 2.6 by using truth tables.

2.8 (i) Show that $\overline{A.B}$ is the dual of $\overline{A} + \overline{B}$.

    (ii) What is the dual of the following circuit?

$$F = \overline{A}.\overline{B} + \overline{A} + C$$

2.9 Convert the equation $F = A \oplus (B \oplus C)$ into a form that requires only AND, OR and NOT gates and obtain its truth table.

2.10 A company is controlled by managing director A, financial director B and two elected members of the board C and D. A needs the support of one other and B needs the support of two others in order to make a decision. Obtain a truth table for the decision making strategies and identify the voting when the decision goes

    (i) against A

    (ii) against B.

2.11 Specify the truth table for a logic system which indicates whether a 4-bit binary number is exactly divisible by 2 or 3.

2.12 A 'two in four' error detecting code is to be used to encode base 6 numbers. Specify a code table for the digits and obtain the truth table of the error detector for this data.

# 3 Combinational logic design

**Objectives**

☐ To define logic levels in an electronic circuit.
☐ To investigate canonical logic forms.
☐ To examine the Karnaugh map method of logic minimization.
☐ To design NAND logic systems and NOR logic systems.
☐ To define and utilize 'don't care' conditions.
☐ To investigate electronic 'hazards'.
☐ To present a tabular method of logic minimization that is suitable for computer implementation.
☐ To examine the principles of cellular logic.

This chapter introduces the principles of combinational logic design at the gate level. We start with the definition of logic levels and proceed with the design and optimization of combinational logic functions that can be built out of discrete gates. The final section of the chapter deals with cellular logic, where design principles are required to produce an individual cell, which can then be repeated, using integrated circuit methods, to produce complex devices.

### Assignment of logic levels

Before designing a logic system, a method of representing binary data within an electronic system must be defined. Each input and output can only have two states, which represent the binary symbols 0 and 1. Consider an electronic circuit having two inputs, A and B, and a single output, F. An example of input/output behaviour is given in Table 3.1 where the allowable voltage states are 0 V and 5 V.

**Table 3.1 Electrical behaviour of a binary system, showing the voltages at the inputs and outputs**

| Inputs | | Output |
|---|---|---|
| A | B | F |
| 0 V | 0 V | 0 V |
| 0 V | 5 V | 0 V |
| 5 V | 0 V | 0 V |
| 5 V | 5 V | 5 V |

If positive coding is used where the symbol 1 is assigned to the positive or higher voltage and 0 to the negative or lower voltage, the truth table for the circuit becomes

| A | B | F |
|---|---|---|
| 0 | 0 | 0 |
| 0 | 1 | 0 |
| 1 | 0 | 0 |
| 1 | 1 | 1 |

and it performs the AND function $F = A.B$.

Using negative coding where $0\,V \equiv 1$ and $5\,V \equiv 0$ the truth table now becomes

| A | B | F |
|---|---|---|
| 0 | 0 | 0 |
| 0 | 1 | 1 |
| 1 | 0 | 1 |
| 1 | 1 | 1 |

and the same electronic system now performs the OR function $F = A + B$. This is an important and unexpected result. Inverting the binary state assignments of an electronic system does *not* result in the inversion of the circuit's function. A circuit that with positive logic coding performs the AND function will become an OR gate with negative logic coding. It does *not* become a NAND gate. The implications of this behaviour are far reaching. Manufacturers must specify, and users adhere to the logic coding of digital devices. If the coding is changed, a logic system will perform a completely different function, although its circuit and electronic behaviour have not been altered. Most manufacturers use positive coding, although negative, and in some cases mixed, coding can be used.

NAND is the inverse of AND.

---

If a circuit performs the NAND function with positive logic coding, what function does it perform with negative logic coding?

**Worked Example 3.1**

*Solution*  Assume the positive logic coding levels are $+5\,V \equiv 1_2$ and $0\,V \equiv 0_2$. The voltage table for the circuit $F = A.B$ is

| A | B | F |
|---|---|---|
| 0 | 0 | 5 |
| 0 | 5 | 5 |
| 5 | 0 | 5 |
| 5 | 5 | 0 |

Now, redefining the voltage levels for negative logic coding such that $0\,V \equiv 1_2$ and $+5\,V \equiv 0_2$, the following truth table is obtained:

The first entry in the truth table corresponds to the last entry in the voltage table, etc.

| A | B | F |
|---|---|---|
| 0 | 0 | 1 |
| 0 | 1 | 0 |
| 1 | 0 | 0 |
| 1 | 1 | 0 |

This is the truth table of a NOR function.

---

An electronic circuit has three inputs, A, B and C, and one output, F, all of which are restricted to either $-5\,V$ or $+15\,V$. Its electrical behaviour is

| A | B | C | F |
|---|---|---|---|
| $-5\,V$ | $-5\,V$ | $-5\,V$ | $-5\,V$ |
| $-5\,V$ | $-5\,V$ | $+15\,V$ | $+15\,V$ |
| $-5\,V$ | $+15\,V$ | $-5\,V$ | $-5\,V$ |
| $-5\,V$ | $+15\,V$ | $+15\,V$ | $-5\,V$ |
| $+15\,V$ | $-5\,V$ | $-5\,V$ | $-5\,V$ |
| $+15\,V$ | $-5\,V$ | $+15\,V$ | $-5\,V$ |
| $+15\,V$ | $+15\,V$ | $-5\,V$ | $+15\,V$ |
| $+15\,V$ | $+15\,V$ | $+15\,V$ | $+15\,V$ |

What logic function does the circuit perform with
(a) positive coding
(b) negative coding?

**Specification of a combinational logic circuit**

In order to design a combinational circuit the inputs and their corresponding outputs must be identified. A combinational system having more than one output can be considered as several single-output logic circuits operating on common inputs. Once the input/output relationship has been defined, a truth table can be calculated.

Consider a combinational logic system that accepts 3-bit binary numbers and indicates if they are in the range $3_{10}$ to $5_{10}$. Three input lines A, B and C are required together with a single bit output F, where $F = 1$ means the input is within range, otherwise $F = 0$.

The truth table for the system can now be compiled. There are eight possible input values. For each input, the designer must determine whether or not the numerical value of ABC is within the specified range. The truth table for the logic system is given in Table 3.2.

**Table 3.2   Truth table for 3 to 5 range indicator**

| A | B | C | F |
|---|---|---|---|
| 0 | 0 | 0 | 0 |
| 0 | 0 | 1 | 0 |
| 0 | 1 | 0 | 0 |
| 0 | 1 | 1 | 1 |
| 1 | 0 | 0 | 1 |
| 1 | 0 | 1 | 1 |
| 1 | 1 | 0 | 0 |
| 1 | 1 | 1 | 0 |

## The 1st canonical form

The 1st canonical form of a combinational logic function can be obtained from the truth table. It consists of a set of minterms that are AND functions of the input variables or their inversions. The outputs of the AND operations are ORed together to give a single output. In a 1st canonical form circuit the ANDing is always carried out on the input data before the OR operation.

Alternative name for 1st canonical form is sum of products or SOP.

The 1st canonical form can be obtained from a propositional description of a truth table. Taking the truth table 3.2 as an example, we have

F is 1   iff A is 0 and B is 1 and C is 1
       or A is 1 and B is 0 and C is 0
       or A is 1 and B is 0 and C is 1

which can be expressed as the Boolean equation

$$F = \overline{A}BC + A\overline{B}\,\overline{C} + A\overline{B}C \tag{3.1}$$

whose circuit diagram is given in Fig. 3.1.

Fig. 3.1   A 1st canonical form circuit ($-\!\!\circ$ on inputs indicates an inversion).

The rule for obtaining the 1st canonical form from a truth table is as follows:

> For every entry with an output at 1, create an AND of the input variables. The variable is used if its value is 1, or its inverse, if 0. The outputs of the AND gates are then ORed together to obtain the final output.

If input A is 1, use A. If input A is 0, use $\overline{A}$.

### Shorthand notation for the 1st canonical form

In order to avoid having to write down long Boolean equations for complex systems, the following shorthand notation will be adopted.

For a 1st canonical form function

$$f(ABC) = \overline{A}BC + A\overline{B}\,\overline{C} + A\overline{B}C$$

the variables in the minterms are replaced with their binary values, giving

$$f(ABC) = 011 + 100 + 101$$

Each minterm is regarded as a number and its decimal value obtained

$$f(ABC) = 3 + 4 + 5$$

35

The function is now represented by the set of the decimal numbers. Hence

$$f(ABC) = \Sigma(3, 4, 5)$$

where $\Sigma$ indicates the 1st canonical form.

---

**Worked Example 3.2**

Express $F(ABCD) = \Sigma(3,4,9,10)$ as a Boolean equation.

*Solution*

Minterm  3 is 0011 and represents $\overline{A}\,\overline{B}CD$
Minterm  4 is 0100 and represents $\overline{A}B\overline{C}\overline{D}$
Minterm  9 is 1001 and represents $A\overline{B}\,\overline{C}D$
Minterm 10 is 1010 and represents $A\overline{B}C\overline{D}$

Hence

$$F = \overline{A}\,\overline{B}CD + \overline{A}B\overline{C}\overline{D} + A\overline{B}\,\overline{C}D + A\overline{B}C\overline{D}$$

---

$F = \Sigma(3,4,9,10)$ could be a function of more than four variables. If it is, the number must be specified with the equation.

If the number of variables in the function is not specified, it can be assumed to be the index of the first integer power of 2 greater than the highest decimal assignment in the function set. In Worked Example 3.2, the highest decimal assignment is 10. The first integer power of 2 greater than 10 is $2^4 = 16$. The index is 4; hence there are four variables in the function.

**The 2nd canonical form**

Alternative name is product of sums or POS.

Important. Compare the 1st canonical form structure.

The 2nd canonical form is an alternative hardware structure to the first form. The input variables are ORed together to form maxterms which are then ANDed together to give the final output. The OR operation is always carried out before the AND.

The 2nd canonical form can be obtained from the truth table by defining the function by its $F = 0$ terms. Taking truth table 3.2 we have

F is 0   iff A is 0 and B is 0 and C is 0
          or A is 0 and B is 0 and C is 1
          or A is 0 and B is 1 and C is 0
          or A is 1 and B is 1 and C is 0
          or A is 1 and B is 1 and C is 1

which gives the Boolean equation

Note $\overline{F}$ is $F = 0$.

$$\overline{F} = \overline{A}\,\overline{B}\,\overline{C} + \overline{A}B\overline{C} + \overline{A}B\overline{C} + AB\overline{C} + ABC \tag{3.2}$$

Inverting both sides gives

Equation 3.3 is of the form
$F = \overline{X_1 + X_2 + \ldots + X_n} = \overline{X_1}.\overline{X_2}.\overline{X_n}$

$$F = \overline{\overline{A}\,\overline{B}\,\overline{C} + \overline{A}\,\overline{B}C + \overline{A}B\overline{C} + AB\overline{C} + ABC} \tag{3.3}$$

Applying De Morgan's theorem to RHS gives

$$F = \overline{\overline{A}\,\overline{B}\,\overline{C}}.\overline{\overline{A}\,BC}.\overline{\overline{A}B\overline{C}}.\overline{AB\overline{C}}.\overline{ABC}. \tag{3.4}$$

Applying the second De Morgan theorem to each NAND term in Equation 3.4

36

gives

$$F = (A + B + C).(A + B + \overline{C}).(A + \overline{B} + C).(\overline{A} + \overline{B} + C).(\overline{A} + \overline{B} + \overline{C}) \quad (3.5)$$

Note the 'OR followed by AND' structure.

This is the 2nd canonical form of the logic and its circuit diagram is given in Fig. 3.2.

The 2nd canonical form can be obtained directly from the truth table without having to resort to lengthy algebra. Comparing Equation 3.5 with truth table 3.2 we can identify each maxterm with an entry that gives a zero output, provided we represent each inverted variable in the maxterm with a one and each normal variable with a zero.

For example:

| A | B | C | F |
|---|---|---|---|
| 0 | 0 | 0 | 0 |
| 0 | 0 | 1 | 0 |
| 0 | 1 | 0 | 0 |
| 0 | 1 | 1 | 1 |
| 1 | 0 | 0 | 1 |
| 1 | 0 | 1 | 1 |
| 1 | 1 | 0 | 0 |
| 1 | 1 | 1 | 0 |

Input ABC = 000 gives F = 0 and forms the maxterm $(A+B+C)$
Input ABC = 001 gives F = 0 and forms the maxterm $(A+B+\overline{C})$
and so on

The rule for obtaining the 2nd canonical form is therefore:

For each entry in the truth table with an output at 0, create an OR of the inputs using the normal variable, if its value is 0, or its inverted value, if the input is at 1. The output of the OR gates are ANDed together to give the final output.

Compare this rule with the 1st canonical rule. It is the opposite in every respect.

### Shorthand notation for the 2nd canonical form

The 2nd form can be represented as a set of numbers identifying the maxterms. Normal variables are assigned 0 and inverted variables 1. Thus

Compare with 1st canonical form, where the normal variable implied 1 and the inverted variable implied 0.

$$F = (A + B + C).(A + B + \overline{C}).(A + \overline{B} + C).(\overline{A} + \overline{B} + C).(\overline{A} + \overline{B} + \overline{C})$$

gives binary values     (000)    (001)    (010)    (110)    (111)
and decimal values     0        1       2       6       7

Hence $\Gamma = \Pi(0,1,2,6,7)$ where $\Pi$ indicates the 2nd canonical form.

If the 1st and 2nd canonical circuits are obtained from the same truth table, they may be quite different from each other, but will always produce identical

Fig. 3.2   A 2nd canonical form circuit.

outputs for the same inputs. They both perform identical operations on the input data.

---

**Worked Example 3.3**  Obtain the 1st and 2nd canonical forms for the EXCLUSIVE OR function.

*Solution*  The truth table for EX.OR is

| A | B | F |
|---|---|---|
| 0 | 0 | 0 |
| 0 | 1 | 1 |
| 1 | 0 | 1 |
| 1 | 1 | 0 |

1st canonical form minterms are

F = 1  when A is 0 and B is 1,  giving $\overline{A}B$
       or A is 1 and B is 0,      giving $A\overline{B}$

ORing the minterms gives

$$F = \overline{A}B + A\overline{B}$$

For the 2nd canonical form maxterms

F = 0  when A is 0 and B is 0,  giving $(A + B)$
       also A is 1 and B is 1,   giving $(\overline{A} + \overline{B})$

ANDing the maxterms gives

$$F = (A + B).(\overline{A} + \overline{B})$$

(Note also that $\overline{A}B + A\overline{B} = (A + B)(\overline{A} + \overline{B})$.)

---

**Conversion between canonical forms**

If the set representation is used, logic functions can readily be converted between 1st and 2nd canonical forms. The set representing the minterms must contain all inputs except for those that are maxterms and vice versa. So for a three-variabled function, the set of all possible inputs is

$$\mathbf{I} = (0,1,2,3,4,5,6,7) \tag{3.6}$$

If the minterm set is $\mathbf{m}$, then the maxterm set $\mathbf{M}$ is

$$\mathbf{M} = \mathbf{I} - \mathbf{m} \tag{3.7}$$

Therefore if $\mathbf{m} = (0,5,6,7)$ then $\mathbf{M}$ would be $(1,2,3,4)$, so the 1st canonical form is

$$F = \Sigma(0,5,6,7)$$

and its corresponding 2nd canonical form is

$$F = \Pi(1,2,3,4)$$

---

If the 1st canonical form of a logic function is

$$F = \overline{A}\,\overline{B}\,\overline{C} + \overline{A}\,BC + A\overline{B}C + ABC$$

what is the equation of the 2nd canonical form?

*Solution*   Given $F = \Sigma(0,1,5,7)$. Hence

$$\mathbf{m} = (0,1,5,7)$$

But F is a function of three variables

$$\therefore \quad \mathbf{I} = (0,1,2,3,4,5,6,7)$$

So

$$\begin{aligned}
\mathbf{M} &= \mathbf{I} - \mathbf{m} \\
&= (0,1,2,3,4,5,6,7) - (0,1,5,7) \\
&= (2,3,4,6) \\
\therefore \quad F &= \Pi(2,3,4,6) \\
&= (A + \overline{B} + C).(A + \overline{B} + \overline{C}).(\overline{A} + B + C).(\overline{A} + \overline{B} + C)
\end{aligned}$$

Remember. In the 2nd canonical form $A \to 0$ and $\overline{A} \to 1$.

---

### Minimal canonical forms

The canonical forms obtained from the truth table can, in most cases, be simplified or minimized while the AND/OR structure of the 1st form or the OR/AND of the 2nd form are still maintained. The minimization theorems from Boolean algebra could be used directly on the logic equations, but this is a somewhat open-ended exercise, and it is quite often difficult to tell when the simplest version has been obtained.

A graphical method of minimizing logic functions was devised by Karnaugh in 1953. It is based on the minimization theorems and Venn diagrams but also guarantees that the circuit is in its simplest form.

### The Karnaugh map

Also known as a K-map.

Consider a Venn diagram containing the binary variable A, occupying half the universe.

The areas must overlap as A and B are not disjoint.

Now, if a second variable B is introduced, which partitions the universe horizontally, we get

Replacing the shaded areas with labels on the axes, we obtain a four-celled K-map, and each cell can contain the output of a logic function when its inputs have the values of the cell co-ordinates. A four-celled K-map can represent any two-variabled logic function. In the K-map of an AND function, a one is placed in the cell corresponding to the intersection of areas A and B on the Venn diagram, whereas the OR function is the union of these areas (see Fig. 3.3).

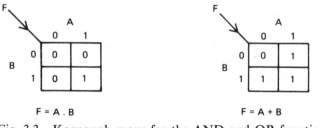

Fig. 3.3 Karnaugh maps for the AND and OR functions.

Karnaugh maps for three- and four-variabled functions can be drawn in two dimensions. Three-dimensional maps can represent functions of up to six variables. As the areas representing the individual variables must overlap, each axis must be labelled in Gray code and cannot be extended to more than two variables. The Gray code labelling is 00, 01, 11, 10 for adjacent rows or columns.

This is not pure binary.

Maps for three- and four-variabled functions are shown in Fig. 3.4. Each cell corresponds to one input state on the truth table and can be labelled with the appropriate output. The address of each cell is the value of its co-ordinates,

40

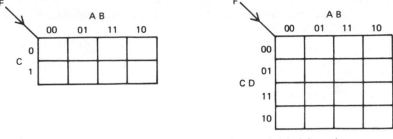

Fig. 3.4    Karnaugh maps for 3- and 4-variabled functions.

expressed as a number and is identical to the decimal assignments used in the set notations for the canonical forms.

---

Obtain K-maps for the functions:

(a) $F = \Sigma(0,2,4,9,11)$

(b) $F = \Pi(3,5,7)$

**Worked Example 3.5**

*Solution*    (a) $F = \Sigma(0,2,4,9,11)$ is a 1st canonical function of four variables, say A, B, C and D. For each minterm a one is entered into the K-map, giving

| | | A B | | | |
|---|---|---|---|---|---|
| | | 00 | 01 | 11 | 10 |
| CD | 00 | $1_0$ | $1_4$ | $0_{12}$ | $0_8$ |
| | 01 | $0_1$ | $0_5$ | $0_{13}$ | $1_9$ |
| | 11 | $0_3$ | $0_7$ | $0_{15}$ | $1_{11}$ |
| | 10 | $1_2$ | $0_6$ | $0_{14}$ | $0_{10}$ |

The address of each cell has been given here as a subscript. Note carefully the order of the numbering due to the Gray coding of the axes.

All remaining cells are set to 0.

(b)  For $F = \Pi(3,5,7)$ we have a 2nd canonical form function of three variables. The maxterm set defines the input conditions where the output is 0. Therefore 0s are entered into the K-map locations defined by the maxterm set and the remaining cells are set to 1.

The K-map is therefore

$\Pi$ means 2nd canonical form. Remember $F = 0$ for a maxterm.

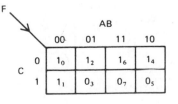

Note the cell addresses.

| | | AB | | | |
|---|---|---|---|---|---|
| | | 00 | 01 | 11 | 10 |
| C | 0 | $1_0$ | $1_2$ | $1_6$ | $1_4$ |
| | 1 | $1_1$ | $0_3$ | $0_7$ | $0_5$ |

*Minimization of canonical forms using Karnaugh maps*

A K-map of a logic function can be used to obtain the simplest version of either the 1st or 2nd canonical form.

Consider the three-variabled logic function $F = \Sigma(1,2,5,6)$. Its K-map is

and full equation

$$F = \overline{A}\,\overline{B}C + \overline{A}B\overline{C} + A\overline{B}C + AB\overline{C} \tag{3.8}$$

Applying the minimization theorem $XY + X\overline{Y} = X$ to minterms 2 and 6, where $X = B\overline{C}$ and $Y = A$, terms $\overline{A}B\overline{C} + AB\overline{C}$ become $B\overline{C}$, which is a considerable simplification.

The minimization theorem is applicable to any minterms on the K-map that occupy adjacent cells. However, by using the map alone, the variables that can be eliminated between a pair of minterms can be detected, without having to resort to algebra.

The procedure is as follows. Neighbouring cells set to 1 are looped together as in Fig. 3.5. K-maps are considered to be continuous at their edges so minterms 1 and 5 are also adjacent and form loop 2.

The function can now be described as being '$F = 1$ if the inputs are in loop 1 or loop 2'. By examining the inputs that define loop 1 we see that B must be 1 and C must be 0, but it does not matter what value A takes, since we still remain within the loop if A is 0 or 1. Loop 1 is therefore independent of A and has the logical function $B.\overline{C}$. Similarly for loop 2, B must be 0 and C must be 1, but A can again be 0 or 1. Loop 2 is therefore $\overline{B}.C$ and the proposition $F = 1$, if in loop 1 or loop 2, may be written as the equation

> This is much simpler than Equation 3.8. However, it still retains the AND/OR structure of a 1st canonical form function.

$$F = B\overline{C} + \overline{B}C \tag{3.9}$$

A similar technique can be used to minimize the 2nd canonical form by looping adjacent 0s on the K-map.

The 2nd canonical form of Equation 3.8 is

$$F = \Pi(0,3,4,7)$$

Fig. 3.5   Looping on a 3-variable K-map (blank cells are assumed to be zero).

and its K-map is

Blank cells are assumed to be set to 1 – omitted for clarity.

Consider maxterms 3 and 7 which are

$$(A + \overline{B} + \overline{C}).(\overline{A} + \overline{B} + \overline{C})$$

This expression is of the form $(X + Y)(X + \overline{Y}) = X$ where $X = (\overline{B} + \overline{C})$ and $Y = A$. The maxterms can therefore be simplified to

See minimization theorems.

$$(\overline{B} + \overline{C})$$

By looping adjacent cells at 0, an input variable can be eliminated. In loop 1, B must be 1 and C must be 1 but A can be either 0 or 1. Loop 1 is therefore independent of A and the simplified maxterm is $(\overline{B} + \overline{C})$.
   The minimal 2nd canonical form for the function is

Remember. In the 2nd canonical form $1 \rightarrow \overline{B}$ and $0 \rightarrow B$.

$$F = (\overline{B} + \overline{C}).(B + C) \tag{3.10}$$

*Rules for looping on a Karnaugh map*

1. Loops must contain $2^n$ adjacent cells set to 1 (or 0 for the 2nd canonical form). A single cell (loop of $2^0$) cannot be simplified. A loop of 2 $(2^1)$ is independent of one variable. A loop of 4 $(2^2)$ is independent of two variables, and in general a loop of $2^n$ is independent of $n$ variables. Therefore to obtain the simplest functions, use the largest possible loops.
2. All cells set to 1 must be covered when specifying the minimal form of the function (0 for the 2nd canonical form).
3. Loops may overlap provided they contain at least one otherwise unlooped cell.
4. Any loop that has all its cells included in other loops, is redundant.
5. Loops must be square or rectangular. Diagonal or L-shaped loops are invalid.
6. There may be different ways of looping a K-map as there is not necessarily a unique minimal form for a given circuit.
7. The edges of a K-map are considered to be adjacent. A loop can leave the top of a K-map and re-enter the bottom and similarly for the sides.

---

Find the minimal 1st and 2nd canonical forms of the function

**Worked Example 3.6**

$$F = \Sigma(3,4,5,6,7,8,10,12,14)$$

*Solution*  The function has four variables and its K-map is

Cells (4, 5, 6 and 7) form a loop of 4 (the largest possible loop for this function). This is loop 1. Cells (8, 10, 12, 14) form another loop of 4 (loop 2) and cells (4, 6, 12, 14) also appear to be a third loop of 4 (loop 3) but all the individual cells are included in either loop 1 or loop 2. Loop 3 is therefore redundant. The final cell to be covered is 3, which combines with 7 to give a loop of 2 (loop 4). The looped K-map is therefore

For clarity when generating a 1st canonical form function from a K-map you may omit cell addresses, and cells at 0 can be left blank.

Loop 1 requires $A = 0$ and $B = 1$ but is independent of C and D.

Loop 1 is therefore $\overline{A}.B$.
Loop 2 requires $A = 1$ and $D = 0$, but is independent of B and C, giving $A\overline{D}$.
Loop 3 is redundant and can be ignored and finally loop 4 has $A = 0$, $C = 1$ and $D = 1$ but is independent of B, giving $\overline{A}CD$.

The minimized version of the 1st canonical form is therefore

$$F = \overline{A}B + A\overline{D} + \overline{A}CD$$

The K-map itself remains the same.

The minimal 2nd canonical form is obtained by looping the zeros on the K-map, giving

Blank cells are assumed to be $F = 1$ and are omitted for clarity.

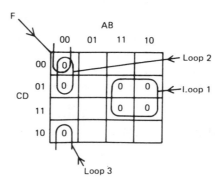

Loop 1 represents $(\overline{A} + \overline{D})$, loop 2 $(A + B + C)$, and loop 3 $(A + B + D)$ giving

$$F = (\overline{A} + \overline{D}).(A + B + C).(A + B + D)$$

---

**NAND logic**

AND, OR and NOT form a universal set of logic gates from which all logic systems can be constructed. It can, however, be shown that these three functions may be carried out with NAND circuits.

A NAND gate will perform the NOT operation if its inputs are common.

Let    $F = \overline{A.B}$
If      $A = B$
Then  $F = \overline{A.A}$
But   $A.A = A$
$\therefore$     $F = \overline{A}$

Hence                     can be replaced by  A —⊐‖о— F

The AND function can be replaced by a NAND gate with its output inverted by a second NAND gate because

$$A.B = \overline{\overline{A.B}}$$

Therefore $\genfrac{}{}{0pt}{}{A}{B}$⊐‖— F is equivalent to $\genfrac{}{}{0pt}{}{A}{B}$⊐‖о—⊐‖о→ F

De Morgan's theorem states that

$$\overline{A.B} = \overline{A} + \overline{B}$$

Inverting all As and Bs gives

$$A + B = \overline{\overline{A}.\overline{B}}$$

This OR function can be implemented with three NAND gates. If the inputs are inverted and then NANDed, the OR function is obtained.

$\genfrac{}{}{0pt}{}{A}{B}$⊐D— F    is equivalent to

NAND is therefore a universal gate – any logic circuit can be constructed solely out of NAND gates.

NAND logic is particularly relevant to 1st canonical form logic systems. Consider the function

$$F = A\overline{B} + BC \tag{3.11}$$

—○ on an input line means the variable is inverted.

The circuit is

The AND and OR gates can be replaced with their NAND equivalents giving

In complex circuits use gate reference numbers.

AND                                    OR

Gates 3 and 5 cancel out, as do 4 and 6.

If every propagation path through the NAND circuit is examined and double inversions removed, the circuit becomes

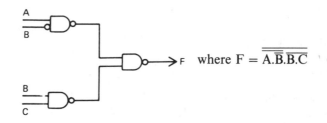

where $F = \overline{\overline{\overline{A.B}.\overline{B.C}}}$

By comparing the final NAND circuit with the original AND/OR version, the reader can see that the structure and interconnections are identical. The only difference between them is that every gate in the AND/OR version has been replaced with a NAND gate.

There is, however, an exception to this behaviour. If the 1st canonical form is not completely two level and any input does not pass through an AND gate, it must be inverted before the AND/OR gates are replaced with NANDS.

Take, for example, the function

$$F = A + BC$$

In a two-level 1st canonical form circuit, each input variable or its inverse first passes through an AND gate (the 1st level of logic) whose output is then input to an OR gate (the 2nd level).

which is a very simple 1st canonical form, where variable A goes directly to the output OR gate. Its circuit is

Direct replacement of the AND/OR gates with the NAND equivalent circuits gives

which simplifies to

where $F = \overline{\overline{A}.\overline{B.C}}$.

The A input must be inverted (with a NAND gate) before the AND/OR gates are replaced with NANDs, thereby satisfying De Morgan's theorem, which, when applied to $F = A + BC$, gives

$$F = \overline{\overline{A}.\overline{B.C}}$$

The rule to obtain the NAND form is as follows. The NAND form of a combinational logic system can be obtained from the minimal 1st canonical form by replacing every gate with a NAND gate provided the original circuit is two level (AND/OR) throughout. If any variable or its inverse only passes through the output OR gate, an additional input inverter must be introduced into that line before all the gates are replaced with NANDs.

## NOR logic

NOR logic is the dual of NAND logic. AND, OR, and NOT functions can be performed with NOR gates only. The equivalent circuits are

Remember NAND form is derived from 1st canonical form. NOR form is derived from 2nd canonical form.

The NOR form can be obtained from the minimal 2nd canonical form by replacing the OR/AND gates with NORs. If, however, a variable or its inverse only passes through the output AND gate, an additional inverter must be introduced on that input line before replacing the gates with NORs.

### 'Don't care' conditions

In some logic systems certain combinations of binary inputs may never occur. If they can be identified their corresponding outputs can be regarded as 'don't cares'. The designer is free to set 'don't care' outputs to either 0 or 1 and should choose the value which gives the simpler logic system.

Assume BCD is 8421 weighted. See Chapter 1.

For example, in BCD coding, where decimal digits are represented in pure binary, the numbers $1010_2$ to $1111_2$ are never used; therefore the outputs of any logic system being driven by BCD encoded data would be 'don't cares' for these inputs.

---

**Worked Example 3.7**

Obtain the minimal NAND and NOR logic functions to drive the top horizontal bar of a seven-segment display from a BCD input.

(A seven-segment display comprises seven light emitting bars a–g. Each bar lights up when a logical 1 is applied to it. The illuminated bars form the shape of the decimal equivalent of the BCD input.)

*Solution* The logic function $F_a$, to drive bar a, is dependent on four binary inputs ABCD. Bar a must be illuminated and therefore $F_a = 1$ when the input has values 0, 2, 3, 5, 7, 8 and 9. Input values 10 to 15 give rise to output 'don't cares' as, in practice, these inputs will never occur.

The K-maps and truth table for $F_a$ are

d stands for 'don't care'.

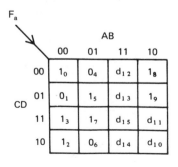

| A | B | C | D | $F_a$ |   | A | B | C | D | $F_a$ |
|---|---|---|---|-------|---|---|---|---|---|-------|
| 0 | 0 | 0 | 0 | 1 |   | 1 | 0 | 0 | 0 | 1 |
| 0 | 0 | 0 | 1 | 0 |   | 1 | 0 | 0 | 1 | 1 |
| 0 | 0 | 1 | 0 | 1 |   | 1 | 0 | 1 | 0 | d |
| 0 | 0 | 1 | 1 | 1 |   | 1 | 0 | 1 | 1 | d |
| 0 | 1 | 0 | 0 | 0 |   | 1 | 1 | 0 | 0 | d |
| 0 | 1 | 0 | 1 | 1 |   | 1 | 1 | 0 | 1 | d |
| 0 | 1 | 1 | 0 | 0 |   | 1 | 1 | 1 | 0 | d |
| 0 | 1 | 1 | 1 | 1 |   | 1 | 1 | 1 | 1 | d |

The NAND form is obtained from the 1st canonical form. The K-map looping is

Loop 1s for the 1st canonical form.

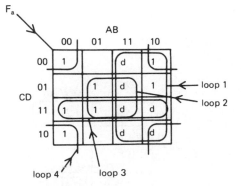

Loop 1 covers cells 8 and 9, but by setting all the 'don't cares' to 1, a loop of eight cells can be drawn. Loop 2 covers cells 5 and 7, which together with 'don't cares' 13 and 14, give a loop of 4. Loop 3 covers cell 3, and loop 4 is a loop of four cells covering all the corners. Therefore

Note in particular loop 4. Side cells are adjacent as are the top and bottom cells.

$$F_a = \text{loop 1 OR loop 2 OR loop 3 OR loop 4}$$
$$= A + BD + CD + \overline{B}\,\overline{D}$$

which can be converted to NAND form by inverting A and replacing each gate with NAND, giving

A only passes through the output OR gate and must therefore be inverted.

$$F_a = \overline{\overline{A}.(\overline{B.D}).(\overline{C.D}).(\overline{\overline{B}.\overline{D}})}$$

Thus the NAND circuit is

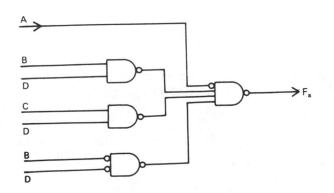

For the NOR version we require the 2nd canonical form. Redrawing the K-map for clarity and omitting the 1s gives

There are three essential 0s to be covered. Cells 4 and 7 are adjacent and form a loop of 4 if 'don't cares' 12 and 14 are set to 0. Cell 1 has no adjacencies set to 0 and remains a loop of 1.

The looped K-map is

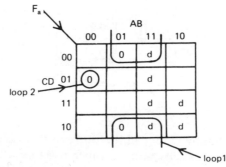

Hence $F_a = (\overline{B} + D).(A + B + C + \overline{D})$.

This is a two-level canonical form and every gate can be replaced with NOR giving

$$F_a = \overline{\overline{(\overline{B} + D)} + \overline{(A + B + C + \overline{D})}}$$

The NOR circuit is

Both circuits also require two input inverters.

Comparing the NAND and NOR forms we see that the latter, requiring three gates, is simpler than the NAND version, which needs four gates.

**Exercise 3.2**

Complete the design of the seven-segment display driver and obtain the simplest version of functions $F_b$ to $F_g$ which drive bars b to g. All are functions of the same input variables A, B, C and D.

## Hazards in combinational logic

Practical electronic logic circuits are not ideal devices. They require a finite time to operate and consequently introduce delays into the propagation of information. These delays are generally no more than a few microseconds and for very high speed logic, may be of the order of nanoseconds, but nevertheless they may invalidate the laws of Boolean algebra and cause errors or hazards in the logic state.

1 nanosecond equals $10^{-9}$ seconds.

Consider the Boolean theorem

$$A.\overline{A} = 0 \qquad\qquad (3.12)$$

The output should be permanently set to 0, regardless of the binary value of A. The logic circuit for this equation is

Let X be the output of inverter gate 1. If A is switched from 0 to 1, the direct line to the AND gate conveys this change immediately. The output (X) of the inverter, which also has A as input, will switch from 1 to 0, but it will take a finite time d$t$, to operate. The timing waveforms for the circuit are shown in Fig. 3.6. The output is not always at logical 0 as one would expect. At certain times the circuit contravenes Boolean algebra and $A.\overline{A} = 1$. This is known as a hazard, and results in a circuit malfunction.

The AND gate is assumed to have zero delay.

Hazards are more likely to occur in multilevel logic circuits as the probability that different propagation paths through the system will have unequal delays is greater. For this reason, canonical forms where there are, at most, three levels of logic, including any input inverters, are preferable. Canonical circuits are, of course, not hazard free, but the hazards are relatively simple both to detect and eliminate.

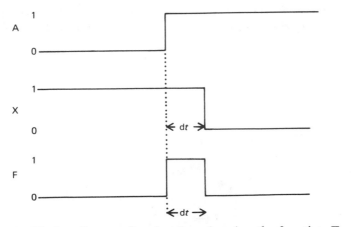

Fig. 3.6   Timing diagram for circuit performing the function $F = A.\overline{A}$.

*Elimination of hazards*

There are three ways of eliminating hazards in combinational circuits. The first method is quite simply to wait. The correct output will always occur once the hazards have passed. If the maximum number of gates in any propagation path is $n$ and the delay of each gate is $dt$, the output of a combinational logic circuit will always be valid at time $T$, after the input data has been applied, where

In practice $T$ never need be greater than a few microseconds.

$$T > n\,dt \tag{3.13}$$

This method is not suitable for combinational circuits whose outputs are used to drive a sequential system, because the incorrect outputs that occur when the hazards are present will become part of the input sequence to the sequential logic.

The second, rather *ad hoc*, method of eliminating hazards is to try to balance out delays by using delay gate arrangements. An AND gate with common inputs will have no effect on the value of the data, but it will present a delay. Hence delays can be introduced deliberately into the propagation paths in a circuit to make them equal to the longest propagation delay.

Some delaying gate arrangements are given in Fig. 3.7.

Fig. 3.7 Some delaying gate arrangements.

A more rigorous technique for overcoming hazards involves the use of K-maps. A hazard will always occur when switching between adjacent cells on a K-map that are unlooped. The hazard can be removed by looping the adjacent cells even if it involves introducing an otherwise redundant term into the function.

---

**Worked Example 3.8**

Identify and eliminate the hazard in the minimal 1st canonical form of the following function:

$$F = \Sigma(1,3,6,7)$$

*Solution*  The K-map is

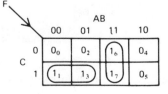

The minimal 1st canonical form is

$$F = AB + \overline{A}C$$

and the circuit

A hazard will occur when moving from cells 7 to 3. Inputs B and C are both at 1 and A switches from 1 to 0. There is an inverter in the propagation path $A \rightarrow X \rightarrow Z \rightarrow F$ and the resulting delay will cause F to switch momentarily to 0, whereas according to the algebra, it should remain at 1. The hazard may be removed by including the redundant loop (3,7) in the covering of the K-map. The circuit will then contain the term BC, which is independent of A and will hold the output F at 1, while A is switching.

The reader should verify this using a timing diagram.

The hazard-free circuit is therefore

$$F = AB + \overline{A}C + BC$$

A K-map can be used to detect and eliminate all hazards arising from the switching of any one variable. Hazards can also arise if two or more variables switch instantaneously. However the likelihood of this happening with independent variables is negligible and may be disregarded.

## Map-entered variables

The standard K-map method of logic minimization is very useful when relatively small digital systems are designed. It is a very elegant 'pencil and paper' technique that avoids algebraic manipulations. If the number of variables is extended to five or six, the map becomes three-dimensional and increasingly complex because of the potential number of overlapping loops. Beyond six, the method breaks down. Furthermore, standard K-maps rely on the designer recognizing the largest possible loops and this human observation process is difficult to automate. Hence, the method cannot easily be programmed into a computer.

In three dimensions the loops become boxes containing $2^n$ cells.

The map-entered variable K-map enables mapping techniques to be extended to functions of more than six variables. The contents of the cells of a standard K-map have so far been restricted to the logical constants 0 and 1. A logic expression can, in the case of a three-variable function, be expressed in the general form

$$F(ABC) = \overline{A}\,\overline{B}\,\overline{C}(f_0) + \overline{A}\,\overline{B}C(f_1) + \overline{A}B\overline{C}(f_2) + \overline{A}BC(f_3)$$
$$+ A\overline{B}\,\overline{C}(f_4) + A\overline{B}C(f_5) + AB\overline{C}(f_6) + ABC(f_7)$$

Universal expansion contains all possible combinations of the variables and their inversions.

where $f_0$ to $f_7$ must take on logical values 0 or 1 depending on the specification of the logic system. These values are then entered into the appropriate cells of the K-map.

The general equation of the three-variable function can also be expanded universally about two variables as follows:

$$F(ABC) = \overline{A}\,\overline{B}(f_0^1) + \overline{A}B(f_1^1) + A\overline{B}(f_2^1) + AB(f_3^1)$$

where $f_i'$, which is sometimes referred to as a residue, is now a function of the logical constants 0 and 1 and the third variable C. This function can now be plotted on a two-variable K-map with axes A and B, and the appropriate residues as cell contents. These are known as the map-entered variables.

Consider the function

$$F(ABC) = \Sigma(1,2,3,5)$$

Expanding this function about three variables gives

The terms ANDed with 0 residues will set F to 0.

$$F(ABC) = \overline{A}\,\overline{B}\,\overline{C}(0) + \overline{A}\,\overline{B}C(1) + \overline{A}B\overline{C}(1) + \overline{A}BC(1)$$
$$+ A\overline{B}\,\overline{C}(0) + A\overline{B}C(1) + AB\overline{C}(0) + ABC(0)$$

The K-map for this function is

In general a *k*-variabled function can be expanded about *l* variables leaving residues of *m* variables and $k = l + m$.

If the same expression is now expanded about only two variables, say A and B, the following equation is obtained:

$$F(ABC) = \overline{A}\,\overline{B}(C) + \overline{A}B(1) + A\overline{B}(C) + AB(0)$$

The logic function in this form can now be plotted on a two-variable K-map with C as a map-entered variable giving

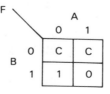

## Minimizing MEV maps

An MEV map can be minimized in two stages.

Step 1 involves setting all the MEVs to 0, giving a conventional K-map which is then minimized. In the above example, setting $C = 0$ we obtain

MEV is map-entered variable.

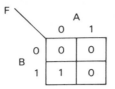

which cannot be reduced and gives the term

$$F_1 = \overline{A}.B$$

Step 2 involves taking the MEV map and changing all the cells set at 1 to 'don't cares'. This is because these cells have already been covered in step 1, and also they are functions of the MEVs and can therefore be combined with other cells containing MEVs.

In this example, you may consider 1 as $C + \overline{C}$ which may be combined with the Cs on the K-map if required. Hence the 'don't care' conditions.

before looping          after looping

Looping is carried out in the normal way except that loops must now contain cells with identical MEVs. The two cells containing the variable C can be looped together and the result in turn is the essential coordinates of the loop (in this case $\overline{B}$) ANDed with the contents of the loop (C) giving $\overline{B}.C$. Don't care cells can be either used or ignored in order to give the best cover of the MEV cells. In this example the don't care terms cannot be used to give an improved cover, and therefore have been disregarded.

The final function is the logical OR of the sub-functions $F_1$ and $F_2$ obtained from the two steps, giving

$$F = \overline{A}B + \overline{B}C$$

The reader can check that this function can be obtained from the original three-variable K-mp. MEV K-maps can be used to extend the K-map minimization techniques to functions containing a larger number of variables than six or to reduce five- and six-variable functions for processing on two-dimensional (four variables or less) maps, and thereby reducing the amount of graphical processing required. It is inadvisable, however, to reduce the size of the K-map too much, because as the map gets smaller the MEVs can become complex logic functions and optimal looping becomes quite difficult to carry out.

**Worked Example 3.9**

Minimize the function $F = \Sigma(2, 3, 8, 9, 10, 11, 12, 13, 14, 15, 20, 21, 22, 23, 24, 25, 26, 27, 29, 31)$.

Each three-variabled term is derived from $4m$ in terms

$\overline{A}\,\overline{B}\,\overline{C}$ covers 0, 1, 2 and 3. As only 2 and 3 are present in this function the residue is $D\overline{E} + DE = D$. $\overline{A}\,\overline{B}C$ covers 4, 5, 6 and 7 and so on.

*Solution*  Assuming the function depends on five input variables A B C D E, expanding universally about A B C gives

$$F = \overline{A}\,\overline{B}\,\overline{C}(D) + \overline{A}\,\overline{B}C(0) + \overline{A}B\overline{C}(1) + \overline{A}BC(1) + A\overline{B}\,\overline{C}(0)$$
$$A\overline{B}C(1) + AB\overline{C}(1) + ABC(E)$$

The MEV K-map is

Step 1, processing cells set to 1, gives

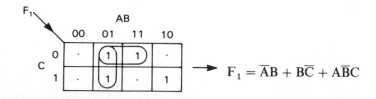

$$F_1 = \overline{A}B + B\overline{C} + A\overline{B}C$$

D and E are MEVs. d is 'don't care'.

Step 2, processing MEVs, gives

$$F_2 = \overline{A}\,\overline{C}D + DE$$

Hence the complete function is

$$\begin{aligned} F &= F_1 + F_2 \\ &= \overline{A}B + B\overline{C} + A\overline{B}C + \overline{A}\,\overline{C}D + DE \end{aligned}$$

and it is in its minimal 1st canonical form.

**Tabular method of minimization**

The Quine–McCluskey algorithm involves the systematic and exhaustive reduction of a Boolean expression. It can handle any number of variables and is easily programmed into a digital computer, although minimization of simple functions can still be carried out by hand.

The starting point of the algorithm is the full list of minterms, if the minimal 1st canonical form is required. Let

Use maxterms for minimal 2nd canonical form.

$$F = \Sigma(0,4,10,11,12,13,14,15)$$

The first step in the algorithm is to group the minterms together according to the number of 1s contained in each input. F is a function of four variables, say W, X, Y and Z; therefore list 1 will have five groups–terms with no 1s, terms containing one 1, two 1s and so on.

*List 1*

| m | WXYZ | m | WXYZ | m | WXYZ | m | WXYZ | m | WXYZ |
|---|------|---|------|---|------|---|------|---|------|
| 0 | 0000 | 4 | 0100 | 10 | 1010 | 11 | 1011 | 15 | 1111 |
|   |      |   |      | 12 | 1100 | 13 | 1101 |   |      |
|   |      |   |      |   |      | 14 | 1110 |   |      |
| | 0 × 1s | | 1 × 1 | | 2 × 1s | | 3 × 1s | | 4 × 1s |

*m* identifies the value of the minterm.

List 2 is obtained by comparing all pairs of entries in adjacent groups in list 1. If a pair differs by 1 bit only, they are combined and the differing variable replaced with a d (for 'don't care'). List 2 is also grouped according to the number of 1s in the pairs, excluding the 'don't care' condition.

This process is equivalent to applying the theorem $AB + A\overline{B} = A$ to every pair of minterms. A can be a single variable or a logic function.

*List 2*

| m | WXYZ | m | WXYZ | m | WXYZ | m | WXYZ |
|---|------|---|------|---|------|---|------|
| 0,4 | 0d00 | 4,12 | d100 | 10,11 | 101d | 11,15 | 1d11 |
|   |      |   |      | 10,14 | 1d10 | 13,15 | 11d1 |
|   |      |   |      | 12,13 | 110d | 14,15 | 111d |
|   |      |   |      | 12,14 | 11d0 |   |      |
| | 0 × 1s | | 1 × 1 | | 2 × 1s | | 3 × 1s |

All pairs of minterms are equivalent to all possible loops of 2 on a K-map of the function.

Lists 1 and 2 are compared to see if every minterm in list 1 is carried through and appears somewhere in list 2. If a term does not carry through, it cannot be combined further. It is an implicant of the function. In this example every term in list 1 carries through to list 2.

An implicant is formed when minterms are combined together. If a minterm cannot be combined it is also an implicant.

List 3 is formed by combining entries in adjacent groups in list 2. Entries may be combined if they differ by one variable and have the same variable eliminated – the 'don't cares' must be common.

*List 3*

| m | W | X | Y | Z |
|---|---|---|---|---|
| 10,11,14,15 | 1 | d | 1 | d |
| 12,13,14,15 | 1 | 1 | d | d |
| | 2 × 1s | | | |

The order of the minterms is unimportant. Term 10, 14, 11, 15 is identical to 10, 11, 14, 15 and is therefore omitted.

The implicants in list 2 which do not carry through to list 3 are (0,4) and (4,12).

Subsequent lists are calculated until no further combination of terms is possible. In this example, list 3 is the final list and terms (10,11,14,15) and (12,13,14,15) are also implicants of the function.

The set of implicants covers the logic function, but like the loops on a K-map, some terms may be redundant.

### Prime implicant table

In order to ascertain the essential or prime implicants, the redundant terms must be identified. This can be done on a prime implicant table, which has columns labelled with the original minterms of the function and rows with the implicants. A flag is inserted at the intersection of rows and columns if the minterm (column label) is contained in the implicant (row label).

We have already identified the implicants (0,4), (4,12), (10,11,14,15) and (12,13,14,15) in the function $F = \Sigma(0,4,10,11,12,13,14,15)$. Its prime implicant table is given in Table 3.3.

**Table 3.3  Prime implicant table for** $F = \Sigma(0,4,10,11,12,13,14,15)$

| Implicants | | Original minterms | | | | | | | |
|---|---|---|---|---|---|---|---|---|---|
| | | 0 | 4 | 10 | 11 | 12 | 13 | 14 | 15 |
| (0,4) | (0d00) | * | * | | | | | | |
| (4,12) | (d100) | | * | | | * | | | |
| (10,11,14,15) | (1d1d) | | | * | * | | | * | * |
| (12,13,14,15) | (11dd) | | | | | * | * | * | * |

By scanning the table, the columns containing only one asterisk identify the essential prime implicants: (0,4) is needed to cover minterm 0 (it also covers 4); (10,11,14,15) to cover 10 and 11; (12,13,14,15) to cover 13. These three prime implicants cover all the original minterms and (4,12) is therefore redundant.

The logic function can be obtained from the binary values of the essential prime implicants giving

From list 2, (0,4) represents
WXYZ = 0d00 and can therefore
be expressed as $\overline{W}\,\overline{Y}\,\overline{Z}$.

$$F = \overline{W}\,\overline{Y}\,\overline{Z} + WY + WX$$

The Quine–McCluskey method, although longer than the K-map minimization, can easily be programmed into a computer and may be extended to any number of variables. (The reader should minimize the previous function on a K-map and show that it gives the same result.)

**Exercise 3.3**   Using the Quine–McCluskey algorithm minimize the following functions
(i) $F = \Sigma(3,4,5,7,9,13,14,15)$

(ii) $F = \Sigma(1,2,3,6,8,9,10,11,18,21,22,24,25,26,27)$
(If you have a computer, you may wish to write a program to solve (ii).)


**Cellular logic**

In some logic systems, particularly where there are a large number of input variables, the canonical forms become increasingly complex and unwieldy. Furthermore, any minor change in the specification of the system often leads to a major redesign of the circuit.

An alternative design approach is to decompose the logic function into a series of identical operations that can be carried out on part of the input data. If a logic cell can be identified and designed, then the complete system may be built up by connecting a number of cells together.

A general cellular circuit is shown in Fig. 3.8. Each cell receives external inputs $I_i$ and generates internal outputs $Q_i$ which are input to an adjacent cell. There may be external outputs at each cell or the output may be derived from the boundary cells, depending on the problem. The direction of flow of information between cells is deliberately not shown in Fig. 3.8 as it is problem dependent. Data flow may be from least to most significant cell or vice versa. In certain other problems it could be either way. The complexity of each cell depends on the number of external inputs it receives, but a trade-off can be made as the larger the number of inputs to a cell, the fewer the cells required in the system.

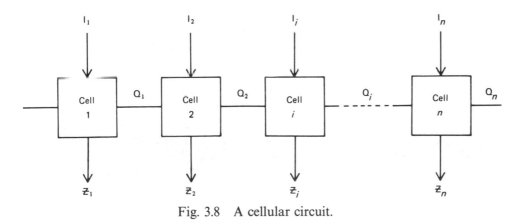

Fig. 3.8   A cellular circuit.


The design of an adder circuit illustrates the essential steps necessary to produce a cellular logic system. Consider a 2-bit adder with inputs $A = A_1A_0$ and $B = B_1B_0$ and outputs $S = S_2S_1S_0$, where

   $S = A + B$

Firstly, for the purpose of comparison, let us obtain the canonical form of the adder. The output consists of three functions $S_2$, $S_1$ and $S_0$ whose truth tables can be evaluated by considering the sum of all possible values of A and B. The K-maps of these truth tables are

Beware. In this equation, + means add

There are 16 different input conditions $A_1 A_0 B_1 B_0$.

59

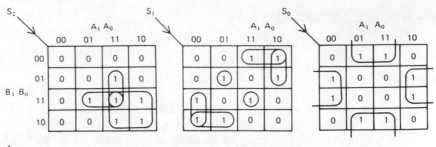

Assume the outputs are functions of all the input variables.

giving

$$S_0 = A_0\overline{B}_0 + \overline{A}_0 B_0 = A_0 \oplus B_0$$
$$S_1 = A_1\overline{B}_1\overline{B}_0 + A_1\overline{A}_0\overline{B}_1 + \overline{A}_1\overline{A}_0 B_1 + \overline{A}_1 B_1\overline{B}_0 + \overline{A}_1 A_0\overline{B}_1 B_0 + A_1 A_0 B_1 B_0$$
$$S_2 = A_1 B_1 + A_0 B_1 B_0 + A_1 A_0 B_0$$

The circuits operate in parallel on common inputs, so the maximum propagation delay will, at most, be three gate delays.

Before designing a cellular circuit for the adder, we must first devise a suitable structure. Let us consider how we, as humans, perform addition. We start with the least significant bits of the input words. Taking 1 bit from each word and any carry from the previous stage, we add them together to produce an output sum bit, and a carry which is input to the next stage of the addition. This procedure is repeated for every binary place in the addition. The adder cell is therefore

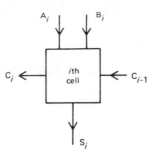

where $A_i$ and $B_i$ are the external inputs, being the $i$th bits of the input words $\mathbf{A}$ and $\mathbf{B}$, respectively. $C_{i-1}$ is the carry-in from the previous cell, $C_i$ is the carry-out to the $(i + 1)$th cell and $S_i$ is an external output forming the $i$th bit of the sum.

Functions $C_i$ and $S_i$ can now be designed. They are combinational logic functions of three input variables $A_i$, $B_i$ and $C_{i-1}$ and represent the sum and carry of the binary addition of 3 bits.

The truth table is

| $A_i$ | $B_i$ | $C_{i-1}$ | $C_i$ | $S_i$ |
|---|---|---|---|---|
| 0 | 0 | 0 | 0 | 0 |
| 0 | 0 | 1 | 0 | 1 |
| 0 | 1 | 0 | 0 | 1 |
| 0 | 1 | 1 | 1 | 0 |
| 1 | 0 | 0 | 0 | 1 |
| 1 | 0 | 1 | 1 | 0 |
| 1 | 1 | 0 | 1 | 0 |
| 1 | 1 | 1 | 1 | 1 |

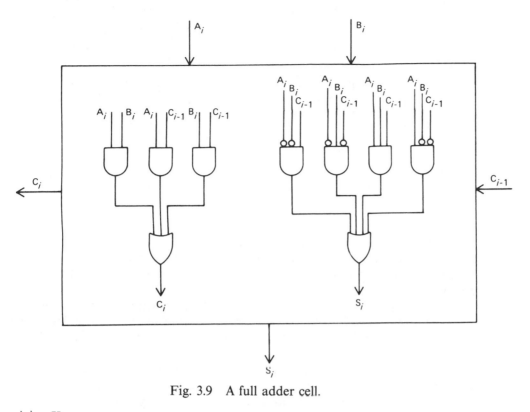

Fig. 3.9   A full adder cell.

giving K-maps

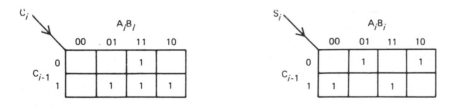

and functions

$$C_i = A_i B_i + A_i C_{i-1} + B_i C_{i-1}$$
$$S_i = \overline{A}_i \overline{B}_i C_{i-1} + \overline{A}_i B_i \overline{C}_{i-1} + A_i B_i C_{i-1} + A_i \overline{B}_i \overline{C}_{i-1}$$

The circuit diagram for the cell is shown in Fig. 3.9. This is often called a full adder and two cells would be required for the 2-bit adder. Some boundary conditions may have to be applied to the end cells. The carry input at cell 0 must be 0 as there cannot be a carry before the least significant bits have been processed. The carry-out from the most significant stage becomes the final sum bit, hence, $S_2 = C_1$. The block diagram of a 2-bit cellular adder is given in Fig. 3.10.

$S_i$ does not simplify as a canonical form. If EX.OR gates are available it can be simplified by algebra to $S_i = A_i \oplus (B_i \oplus C_{i-1})$.

If the cell is designed with only one external input it is known as a half adder. Twice as many cells would be required compared with full adder circuits.

61

Fig. 3.10 A 2-bit adder circuit.

The main advantage of the cellular structure over the canonical form is that the former can be expanded by simply adding on extra cells. A 3-bit adder would need one extra cell whereas a 16-bit adder would require a further 13. The cells are all identical – their functions do not change with the increasing resolution of the adder.

Compared with the canonical approach, a 3-bit adder would require a complete redesign of the function $S_2$ and a new function ($S_3$) of six input variables. In the case of a 16-bit canonical form adder, the most significant sum bit would be an unwieldy function of 32 input variables.

Iterative is an alternative name to cellular.

The cellular or iterative logic structure is extremely important in very large scale integration manufacture. If a complex logic system is made in the form of an integrated circuit, the design process becomes relatively simple if a cell can be devised. The complex system then comprises a large number of repeated cells and this may be carried out photographically on the integrated circuit mask or by using a computer-aided drawing package.

The main disadvantage of the cellular form of logic is its operating time, which is directly proportional to the number of cells in the system. In the canonical form of the adder the maximum delay was $3\,\mathrm{d}t$ and this is independent of the resolution of the adder, as all functions effectively operate in parallel from common inputs.

dt is the delay due to one gate.

In the cellular circuit each cell has a delay of $3\,\mathrm{d}t$. Taking the worst case delay in the adder, where each cell generates a carry, a period of $3n\,\mathrm{d}t$ (where $n$ is the number of cells in the system) must elapse before the output is valid, otherwise hazards may be present on the output.

Cells operate in series – delays are cumulative.

---

**Worked Example 3.10**

Design a cellular circuit to generate the odd parity for 4-bit words and compare it with its equivalent canonical form.

Odd parity – an additional bit in a binary word which makes the total number of bits set to 1, an odd number.

*Solution*  The system structure is as follows:

1. *Data flow.* As parity can be calculated by starting with either the least or most significant bit, the direction of data flow does not matter in this problem.
2. *Cell specification.* Let each cell process 1 bit of the input word and calculate $P_i$ – the 'parity so far'. The final output P is the 'parity so far' after all the inputs have been processed. No external output is required at each cell.

The block diagram for the cell is

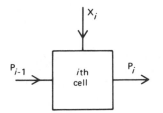

If the parity ($P_{i-1}$) for inputs $X_0$ to $X_{i-1}$ is 0, the number of bits set to 1, between bits $X_0$ to $X_{i-1}$, must be odd. If the current input $X_i$ is 0, the new 'parity so far' ($P_i$) needs to be 0. The parity for all other cell input conditions can also be calculated and the cell truth table obtained, which is

| $P_{i-1}$ | $X_i$ | $P_i$ |
|-----------|-------|-------|
| 0 | 0 | 0 |
| 0 | 1 | 1 |
| 1 | 0 | 1 |
| 1 | 1 | 0 |

P is therefore the EXCLUSIVE OR function

$$P_i = P_{i-1} \oplus X_i$$

so the cell is simply an EX.OR gate.

The complete parity generator is therefore

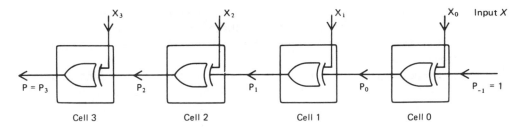

The canonical form of the odd parity generator can be obtained directly from the K-map specification:

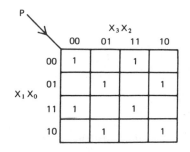

It does not minimize and is therefore

$$P = \overline{X}_3\overline{X}_2\overline{X}_1\overline{X}_0 + \overline{X}_3\overline{X}_2X_1X_0 + \overline{X}_3X_2\overline{X}_1X_0 + \overline{X}_3X_2X_1\overline{X}_0$$
$$+ X_3\overline{X}_2\overline{X}_1X_0 + X_3\overline{X}_2X_1\overline{X}_0 + X_3X_2\overline{X}_1\overline{X}_0 + X_3X_2X_1X_0$$

**Exercise 3.4**   Design a cellular circuit to indicate single errors in a '2 in 5' encoder. (Hint: 2 bits of information must be passed between cells. Let $Q = 00$ be '0 bits set so far', 01 be '1 bit set', 10 be '2 bits set' and 11 be 'more than 2 bits set'. An error is present if output from the final cell is not 10.)

**Summary**

This chapter has been concerned with the fundamental principles of combinational logic design. We have seen that a logic function is dependent on the definition of the logic levels on the inputs and outputs. If the coding is inverted, the dual function is obtained.

Two standard logic forms have been defined. The 1st canonical form has an AND level of logic immediately after the inputs followed by an OR gate, whereas the 2nd form is OR followed by AND. Both the 1st and 2nd canonical forms can be obtained from a truth table and they both perform the same logical operation.

Canonical forms can be minimized. The K-map method is suitable for optimizing small logic systems up to six variables. The Karnaugh method can be extended to functions of more than six variables if map-entered variable (MEV) techniques are used. For very large systems, up to 20 variables, a computerized version of the Quine–McCluskey algorithm can be used.

Every gate in a 1st canonical form circuit can be replaced by a NAND gate, provided the circuit is two layer throughout and every input or its inversion first passes through an AND gate and then via an OR gate to the output. The 2nd canonical form has a similar property and can be converted to NOR logic.

In some logic systems, 'don't care' conditions can be identified on the outputs. The designer is free to set a 'don't care' output to either 0 or 1, and this may influence the minimization.

Electronic logic circuits are imperfect devices. They do not operate instantaneously. The delays they introduce, momentarily, may cause the laws of Boolean algebra to be violated and the circuit to malfunction. Methods of removing these hazards have been examined in this chapter.

Finally, a cellular structure has been proposed that has both advantages and disadvantages compared with the canonical form. A cellular structure is, however, ideally suited to integrated circuit manufacturing techniques.

**Problems**

3.1   The functions of circuits within a logic system with positive logic coding are

$$F_1 = \overline{A}\,\overline{B}\overline{C} + \overline{A}BC + A\overline{B}\overline{C}$$
$$F_2 = A\overline{B}C + AB\overline{C} + ABC$$

What functions do the circuits perform if negative logic coding is used?

3.2 (i) Obtain the minimal 1st canonical forms for the following equations:

$$F_1 = \overline{A}\,\overline{B}\overline{C} + \overline{A}\,\overline{B}C + \overline{A}B\overline{C} + A\overline{B}C$$
$$F_2(ABCD) = \Sigma(3,4,5,7,9,13,14,15)$$
$$F_3(ABC) = \Pi(2,3,4,5,6,7)$$

(ii) Obtain the minimal 2nd canonical form for the above equations.

3.3 Obtain the minimal NAND forms of the above functions:

$$F_1 = \Sigma(3,4,5,6,7,8,10,12,13)$$
$$F_2 = \Sigma(0,1,2,3,4,5,8,9,10,11)$$

3.4 Obtain the minimal NOR forms of the following functions:

$$F_1 = \Sigma(0,2,3,4,6)$$
$$F_2 = \Pi(0,1,2,8,9,12,13,14,15)$$

3.5 Obtain the minimal 1st and 2nd canonical forms of the following function:

$$F = \Sigma(4,5,6,8,10,14)$$
$$\text{'don't cares'} = \Sigma(1,3,7,12,15)$$

3.6 Design a digital system using only the minimum number of NAND gates that will convert 8,4,2,1 (ABCD) into $8,4,-2,-1$ (WXYZ) binary-coded decimal.

3.7 Design a digital system using AND/OR/NOT gates that will convert 4-bit binary (ABCD) into Gray code (WXYZ). What would be the minimum number of gates required if EXCLUSIVE OR gates were made available?

3.8 Design a logic system using the minimum number of NAND gates which will indicate when a 4-bit binary number is exactly divisible by 2 or 3.

3.9 Repeat Problem 3.8, redesigning the system so as to use the minimum number of NOR gates.

3.10 Identify and eliminate the hazards in the following circuit:

$$F = \Sigma(1,2,3,5,12,13,14,15)$$

3.11 Minimize the following function, by means of the Quine–McCluskey algorithm:

$$F = \Sigma(1,4,6,7,12,13,14,15,17,28,29,30,31)$$

3.12 Design a cellular logic system that will accept 8-bit words $A$ and $B$ and indicate whether $A$ is greater than, equal to, or less than $B$.

3.13 Minimize the following function using a three-variable K-map with a map-entered variable:

$$F = \Sigma(1,3,4,5,6,7,11,12,15)$$

3.14 Repeat Problem 3.11 using:
(i) a four-variable K-map and one map-entered variable;
(ii) a three-variable K-map and two map-entered variables.

# 4 Sequential logic fundamentals

Feedback is an essential feature of many natural and artificial systems.

Having examined combinational logic in detail in Chapter 3, we now investigate the effects of applying some feedback to a logic circuit. The feedback creates a sequential system whose output is dependent not only on the present input, but also on all the previous inputs over a given sequence. Another property of a sequential circuit is its ability to give a different output response when the same input is reapplied to the circuit. This behaviour cannot happen in a combinational circuit. In this chapter a simple sequential circuit with a single feedback connection will be used as a prototype from which a range of flip-flops will be developed. The flip-flops can be regarded as the building block of a discrete sequential logic system.

### A sequential logic circuit

A sequential logic circuit, in its simplest form, is a combinational logic system with some feedback connections from the output, providing one or more additional internal inputs. Data is applied to the external inputs, but the user has no direct control over the internal inputs. They can only be changed by the logic circuit itself responding to all its inputs, both external and internal. The value of the internal inputs defines the state of the system.

A sequential logic circuit comprising a combinational system with some feedback is shown in Fig. 4.1. It has a single output Q which is also fed back to form internal input Q′ (the state of the system). The external inputs are S and R and the system can be described by the Boolean equation

$$Q = \overline{R + \overline{(S + Q')}} \tag{4.1}$$

where Q, on being fed back, becomes Q′.

Fig. 4.1  A sequential circuit formed by applying feedback to a combinational circuit.

**Table 4.1**  **Truth table for sequential circuit**
$Q = R + \overline{(S + Q')}$ **with feedback loop open circuit**

| S | R | Q' | Q | Stability | |
|---|---|----|---|-----------|---|
| 0 | 0 | 0 | 0 | Stable | ① |
| 0 | 0 | 1 | 1 | Stable | ② |
| 0 | 1 | 0 | 0 | Stable | ③ |
| 0 | 1 | 1 | 0 | Unstable | |
| 1 | 0 | 0 | 1 | Unstable | |
| 1 | 0 | 1 | 1. | Stable | ④ |
| 1 | 1 | 0 | 0 | Stable | ⑤ |
| 1 | 1 | 1 | 0 | Unstable | |

The output is determined by both the input and state values, so it is possible to obtain a different response if the same inputs are reapplied. The present output at time $t$ is a function of the present inputs and state, but the latter is a function of inputs and state at time $t - 1$. The state at $t - 1$ depends on inputs and state at $t - 2$ and so on. The present output is therefore dependent on the present input and all previous inputs which have been applied to the system. This property characterizes a sequential system.

'Inputs' are the external inputs.

A feedback logic sequential circuit may be analysed by imagining that there is a switch in the feedback loop that isolates Q from Q'. If the switch is open, the circuit becomes a combinational logic system and its truth table can be calculated. The truth table for the feedback circuit in Fig. 4.1 is given in Table 4.1.

Sometimes called a Huffman switch, after D.A. Huffman who first suggested the idea.

When the Huffman switch is closed, Q' is forced to take the value of Q. If Q has the same value as Q', nothing changes and the circuit is said to be STABLE. If, however, Q changes Q', a different set of inputs is now applied to the logic because the state has changed even though the external inputs R and S are held constant. The circuit is therefore UNSTABLE and a new output will be generated. The circuit will continue to switch until a stable state is entered.

If no stable state exists, the circuit will oscillate between unstable states.

The stability of a circuit can be determined from its truth table:

If $Q = Q'$ then the circuit is stable.
If $Q \neq Q'$ the circuit is unstable.

The stability conditions are given in Table 4.1 alongside the output data and can be summarized on a K-map where each stable cell is labelled with a circled number (e.g. ①), giving

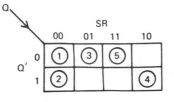

The unstable states may now be examined. Consider SR = 01 and Q' = 1. The output Q becomes 0 and will, when fed back, switch Q' to 0, thereby creating the

67

input conditions for stable state ③. Therefore the circuit state with SRQ' equal to 011 is unstable state 3, which will switch to stable ③. Unstable states 4 and 5 can also be identified. The complete stability map is called a flow table, and is given in Table 4.2.

**Table 4.2  Truth table for the Sequential Circuit $Q = \overline{R + \overline{(S + Q')}}$**

Stable states are circled and their corresponding unstable states have the same reference number, but are uncircled.

Gray coding is convenient if information is being transferred from a K-map to a flow table.

Flow tables are only used to summarize the stability of a circuit. They are not used for minimization; therefore the axes do not necessarily have to be Gray coded, and are not restricted to two variables. The columns are, however, always labelled with the external inputs and the rows with the internal states. Any horizontal motion within the flow table can be brought about by changing the external inputs, whereas the vertical motion is controlled by the stability of the circuit.

A flow table can be used to investigate the behaviour of a circuit. In state ① (Table 4.2) the circuit is stable and S and R are both 0. If S is set to 1, unstable state 4 will be entered. The circuit will then switch itself into stable ④ and both the internal state Q' and the output will become 1. If SR is returned to 00, state ② is entered and the output remains at 1, whereas previously with SR at 00 the output was 0. If R is now set to 1, state ③ is entered and the output becomes 0. The circuit is a 1-bit memory. It will hold the value 1 on its output if S is momentarily set to 1. The memory can be cleared to 0 if the R input becomes 1. This sequential circuit is known as a set–reset flip-flop (SRFF) where S is the set input and R is the reset input.

Flip-flops are sometimes called bistables.

**Worked Example 4.1**  Investigate the stability of the following system and obtain its flow table

$$F_1 = \overline{A}\overline{F}'_1 + \overline{B}\overline{F}'_1 + \overline{A}\overline{F}'_2 + AF'_1F'_2$$
$$F_2 = \overline{A}\overline{B} + \overline{A}F'_2 + \overline{B}F'_1F'_2 + BF'_1F'_2$$

$F_1 \rightarrow F'_1$ reads $F_1$ is fed back to input $F'_1$.

where $F_1 \rightarrow F'_1$ and $F_2 \rightarrow F'_2$.

*Solution*  First obtain the truth tables of $F_1$ and $F_2$. Both are functions of four variables, A, B, $F'_1$ and $F'_2$

| A | B | $F_1'$ | $F_2'$ | $F_1$ | $F_2$ | Stable state |
|---|---|---|---|---|---|---|
| 0 | 0 | 0 | 0 | 1 | 1 | |
| 0 | 0 | 0 | 1 | 1 | 1 | |
| 0 | 0 | 1 | 0 | 1 | 1 | |
| 0 | 0 | 1 | 1 | 1 | 1 | ① |
| 0 | 1 | 0 | 0 | 1 | 0 | |
| 0 | 1 | 0 | 1 | 1 | 1 | |
| 0 | 1 | 1 | 0 | 1 | 0 | ② |
| 0 | 1 | 1 | 1 | 0 | 1 | |
| 1 | 0 | 0 | 0 | 0 | 0 | ③ |
| 1 | 0 | 0 | 1 | 0 | 1 | ④ |
| 1 | 0 | 1 | 0 | 1 | 0 | ⑤ |
| 1 | 0 | 1 | 1 | 1 | 0 | |
| 1 | 1 | 0 | 0 | 0 | 0 | ⑥ |
| 1 | 1 | 0 | 1 | 0 | 0 | |
| 1 | 1 | 1 | 0 | 0 | 0 | |
| 1 | 1 | 1 | 1 | 1 | 1 | ⑦ |

By comparing $F_1'F_2'$ with $F_1F_2$, the stable states can be identified. This circuit is only stable if both $F_1'$ and $F_2'$ are identical to $F_1$ and $F_2$. Seven stable states can be identified and entered onto a flow table.

Now considering the unstable states we have, for inputs $AB=00$ and $F_1'F_2'=00$, outputs $F_1F_2 = 11$. After feedback $F_1'F_2'$ becomes 11 and this is the input condition for stable state ①. Hence $ABF_1'F_2' = 0000$ is unstable state 1.

The complete flow table is

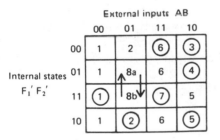

It is left as an exercise for the reader to check for him- or herself all the remaining unstable states. Note in particular the oscillation between unstable states 8a and 8b and that stable state ① has three unstable states associated with it.

Oscillations are indicated by arrows on the flow table.

## Limitations of the set–reset flip-flop

Flow tables represent ideal circuit behaviour. They do not take into account any timing problems. Suppose the SR input data on an SRFF changes from 11 to 00 (see Table 4.2). One would expect the circuit to switch from ⑤ to ① and the

Fig. 4.2    Data values in a SRFF when SR = 11.

Typical value is 1 microsecond.

output to remain at 0. In practice, S and R will not both change at the same instant of time. If S switches before R, the circuit will enter ③ and then ① and the output will remain at 0. If, however, R becomes 0 before S, we leave ⑤ and enter unstable state 4. The circuit will automatically switch to stable ④ and then to ② when S becomes 0. The output is now 1. The behaviour of the circuit is therefore uncertain and depends on whether S or R switches first. It will be sensitive to very small time differences between the two inputs.

The problem would appear to be soluble by electronically synchronizing the inputs. Unfortunately this will reveal another timing problem. Figure 4.2 shows the data values on all gate inputs and outputs when an SRFF is in state ⑤. If S and R are both switched simultaneously to 0, the inputs to both NOR gates will be at 0. The gates will both output 1, and X (the output of gate 1) will be propagated to X' (the input of gate 2), while Q (the output of gate 2) will be fed back to Q' (the input of gate 1). If the propagation times are identical, X' and Q' will become 1 at the same instant. Both gates will have inputs 10 and will output 0. X and Q become 0 again. The switching procedure then repeats itself and the output will oscillate between 0 and 1. In practice the delays will never be exactly the same. If X to X' is shorter than Q to Q' the output will remain at 0 (state ①). If not, state ② is entered and the output becomes 1. The final state of the circuit therefore will depend on the internal propagation delays.

If the flip-flop is regarded as a memory, SR = 11 represents the impossible request to set a single bit to both 1 and 0 at the same time.

The designer can avoid this flip-flop hazard by never allowing S and R both to be 1 at the same time. The problem associated with switching inputs from 11 to 00 will then never occur. If this restriction is enforced, the output, X, of gate 1 then becomes the inverse of the output of gate 2, i.e. $X = \overline{Q}$ for all allowable inputs.

The final truth table of an SRFF is

| S | R | $Q_t$ | $Q_{t+dt}$ |
|---|---|-------|------------|
| 0 | 0 | 0 | 0 |
| 0 | 0 | 1 | 1 |
| 0 | 1 | 0 | 0 |
| 0 | 1 | 1 | 0 |
| 1 | 0 | 0 | 1 |
| 1 | 0 | 1 | 1 |
| 1 | 1 | 0 | d |
| 1 | 1 | 1 | d |

If SR = 11 is not allowed, the corresponding output becomes 'don't care'.

dt is the switching time of the circuits.

where $Q_t$ is the present internal state (equal to the output before the circuit has operated) and $Q_{t+dt}$ is the next state and output after the circuit has responded to the input data SR.

The transition table for the SRFF defines the inputs required to give a specific data transition from present state to next state ($Q_t \rightarrow Q_{t+dt}$). There are four possible transitions, $0 \rightarrow 0$, $0 \rightarrow 1$, $1 \rightarrow 1$ and $1 \rightarrow 0$. By comparing $Q_t$ with $Q_{t+dt}$ in the truth table, we see that the transition $0 \rightarrow 0$ occurs if SR = 00 or 01. S must therefore be 0 but R can be 0 or 1, so R is a 'don't care'. The other transitions can be detected on the truth table and the input requirements are given in the SRFF transition table (Table 4.3).

In this circuit the state and output transitions are identical.

**Table 4.3  Transition table for an SRFF**

| S | R | $Q_t \rightarrow Q_{t+dt}$ |
|---|---|---|
| 0 | d | $0 \rightarrow 0$ |
| 1 | 0 | $0 \rightarrow 1$ |
| d | 0 | $1 \rightarrow 1$ |
| 0 | 1 | $1 \rightarrow 0$ |

The circuit diagram of the SRFF is often drawn as a pair of cross-coupled gates (see Fig. 4.3). The NOR version may be converted to NAND by Boolean algebra. The SRFF is a simple sequential circuit having only one feedback loop. It is asynchronous and operates solely on the input data S and R. No control data is required to determine when the circuit will switch. It operates immediately the inputs are applied. SRFFs form the basis of a family of bistables which have additional circuitry to overcome some of their practical limitations. Flip-flops can be regarded as the elementary building blocks of sequential systems.

Note the inverters on the inputs and outputs of the NAND circuit.

NOR version

NAND version

Symbol

Fig. 4.3   Circuit diagrams of NAND and NOR versions of an SRFF.

**Switch debouncing using a set−reset flip-flop**

One important and direct application of an SRFF is the debouncing of an electro-mechanical switch. A typical switch arrangement that produces a logic output F and its inverse $\overline{F}$ is shown in Fig. 4.4. The F and $\overline{F}$ terminals are connected to earth via resistors $R_1$ and $R_2$ and the toggle contact is at $+5$ V.

Fig. 4.4   A switch circuit to produce a logical constant and its inverse.

When the switch is in the $F = 0$ position, there is no voltage across $R_1$, $F$ is at earth potential (logical 0), 5 V appears across $R_2$ and $\bar{F}$ is at logical 1. While the switch is being operated, there will be a short period of time when the toggle is between the contacts of $F$ and $\bar{F}$ and both outputs will be 0. When the toggle reaches $F$, it will become 1, but the toggle will physically bounce off the $F$ contact, leaving the surface, momentarily causing $F$ to be 0 again. It will then rebound an indeterminant number of times causing a string of 0s and 1s to be output before settling to $F = 1$. The bounce may be eliminated electronically, by connecting the switch output $F$ to the S input, and $\bar{F}$ to the R input of an SRFF. The outputs Q and $\bar{Q}$ will then be the debounced versions of the switch outputs.

Bouncing is a mechanical effect that occurs in all simple switches. The electronics is sensitive to bounces over a very short period of time, typically 1 microsecond.

**Worked Example 4.2**

Show, using data waveforms, that an SRFF eliminates bounce from the switch arrangement in Fig. 4.3.

*Solution*

Bouncing only occurs on the contacts that are 'making'. The 'breaking' contacts always give a clean pulse.

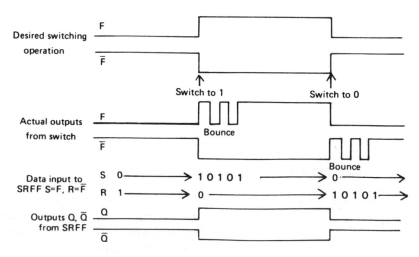

Q and $\overline{Q}$ give the desired bounce-free waveforms because the flip-flop is set (or reset) on the first touching of the contacts. The first bounce causes $S = R = 0$ and this does not change the outputs Q and $\overline{Q}$. The second time the contacts touch, the data will set (or reset) the flip-flop, but it has already been set (or reset) by the very first contact. There is consequently no change in Q and $\overline{Q}$ and the spurious 1s and 0s due to the bouncing are eliminated.

Refer to SRFF truth table.

Flip-flop is set when the switch goes to 1 and reset when switch goes to 0.

Show by means of data waveforms that two cross-coupled NAND gates will eliminate the bounce on a switch which has its toggle contact earthed (logical 0) and dropper resistors connected to $+5\,V$ (logical 1).

**Exercise 4.1**

### The JK flip-flop

The input restrictions on the SRFF may be overcome if a JK flip-flop (JKFF) is used. This bistable consists of an SRFF with additional gating logic on the inputs. The logic diagram of the JKFF is given in Fig. 4.5. The external inputs are J and K where J is ANDed with $\overline{Q}$, the inverse of the output of the SRFF, and drives S. K is ANDed with Q and drives R.

Restrictions. $S = R = 1$ is not allowed.

Fig. 4.5   Simple JK flip-flop.

A third input, the clock, is common to both AND gates. Its function is to control the switching of the bistable and to synchronize the inputs. If the clock is at logical 0, both AND gate outputs to S and R will be 0 and the flip-flop will not therefore change state. The reader can evaluate the behaviour of the JKFF by referring to the SRFF truth table and calculating the logic values passing through the input logic, for all possible values of J and K. In particular, consider the case where both inputs are set to 1. If Q and the clock are at 1, the logic values transmitted through the AND gates are $S = 0$ and $R = 1$. This will reset the SRFF to $Q = 0$ and $\overline{Q} = 1$. These new outputs, when fed back via the input logic, will set SR to 10 and its outputs will become $Q = 1$, $\overline{Q} = 0$. Hence the output of the JKFF will oscillate continuously between 0 and 1 when both J and K are at 1.

In the simple JKFF this oscillation will always happen, whereas with both inputs set to 1 on an SRFF, the final value of the output can be 0 or 1 depending on the internal delays within the circuit.

J and K have no literal meaning.

This input condition is not allowed on an SRFF.

Although both J and K are 1, S and R can never both become 1 at the same time.

## The master–slave principle

Although the input logic in a simple JKFF eliminates some of the timing problems inherent in the SR device, its final state still cannot be determined when $J = K = 1$, as it depends on the length of time the circuit is allowed to oscillate. If the circuit was only allowed to switch once, then the behaviour could be completely specified. Single switching could be achieved by using a very short clock pulse, but this would give rise to technical difficulties. The alternative is to use a master–slave device.

A master–slave JKFF (Fig. 4.6) uses two SRFFs, together with input gating logic. The output of the second, or slave, SRFF is fed back to the input logic of the first SRFF (master). A most important feature is the inverter in the clock line. The clock is directly connected to the master but passes through an inverter before entering the slave. When the clock is high, the inputs switch the master and the feedback from $Q_2$, and $\overline{Q}_2$ prevents $S_1$ and $R_1$ from being simultaneously 1. The slave inputs $S_2$ and $R_2$ remain at 0 owing to the inverter in the clock line inhibiting the output of gates 3 and 4. As the clock switches to 0, data can be input to the slave as its clock, $Cl_2$, now becomes 1. The slave inputs are the outputs of the master, which remain constant because $Cl_1$ is 0 and no new data can enter the master. The output of the slave is determined by the state of the master which is governed by the values of J and K the instant the clock switches from 1 to 0. The master–slave system can therefore be considered as a JKFF which is switched by the data on the inputs at the instant a falling edge occurs on the clock line. Only one change of state Q can occur for each clock pulse.

The truth table for a master–slave JKFF is

Master does not change because $S_1 = R_1 = 0$.

$Q_t$ is often called the PRESENT state and $Q_{t+1}$ the NEXT state.

$t$ is measured in clock pulses.

| J | K | $Q_t$ | $Q_{t+1}$ | Effect on output |
|---|---|---|---|---|
| 0 | 0 | 0 | 0 | no change |
| 0 | 0 | 1 | 1 | |
| 0 | 1 | 0 | 0 | reset to 0 |
| 0 | 1 | 1 | 0 | |
| 1 | 0 | 0 | 1 | set to 1 |
| 1 | 0 | 1 | 1 | |
| 1 | 1 | 0 | 1 | change state |
| 1 | 1 | 1 | 0 | |

where J and K are the external inputs, $Q_t$ is the output (and internal state) before, and $Q_{t+1}$ is the output after, the clock edge has occurred.

The transition table which gives the required inputs for each output transition is

| J | K | $Q_t \to Q_{t+1}$ |
|---|---|---|
| 0 | d | $0 \to 0$ |
| 1 | d | $0 \to 1$ |
| d | 0 | $1 \to 1$ |
| d | 1 | $1 \to 0$ |

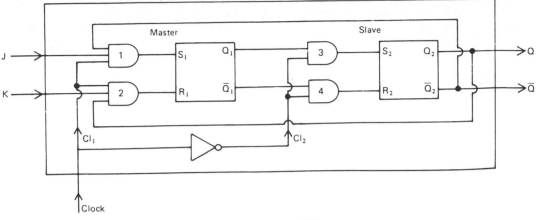

Fig. 4.6   A master–slave JKFF.

Almost all commercially available JKFFs are of the master–slave type and operate on the falling edge of the clock. If any circuit contains JKFFs they should be assumed to be of the master–slave type.

The following data is applied to J, K and the clock of a master–slave JKFF. If its initial state is $Q = 0$, what is the output waveform?

**Worked Example 4.3**

*Solution*

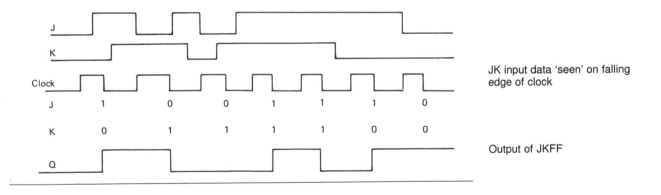

JK input data 'seen' on falling edge of clock

Output of JKFF

Design a master–slave JK flip-flop using NOR gates only.

**Exercise 4.2**

### The trigger flip-flop

Two types of trigger flip-flop (TFF) may be designed by restricting the input conditions on a JKFF. In the asynchronous TFF the inputs to the internal JKFFs

Sometimes called a toggle.

are connected permanently to 1, and data is input via the clock line, as shown in Fig. 4.7. The output changes each time a falling edge occurs on T. There will be one change on Q either $(0 \rightarrow 1)$ or $(1 \rightarrow 0)$ for each complete pulse on T. Typical input and output waveforms are

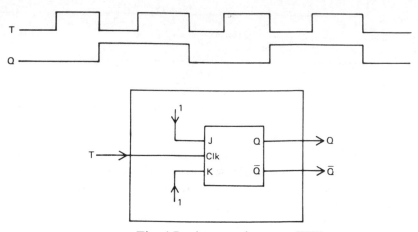

Fig. 4.7   An asynchronous TFF.

The output Q contains exactly half the number of pulses in the input T. As the asynchronous trigger divides the input by 2, it may therefore be used to build up binary counting circuits. Let the output of an asynchronous TFF drive another one as shown in Fig. 4.8. The waveforms on $T_0$, $Q_0$ and $Q_1$ will be

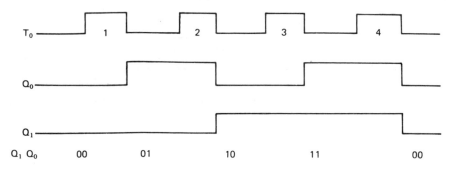

The binary value of $Q_1 Q_0$ immediately after a pulse on $T_0$ is numerically equal to the number of pulses that have occurred on $T_0$. The system is a 2-bit pure binary counter, and its range can quite easily be extended by adding on extra TFFs.

Fig. 4.8   A 2-bit pure binary counter.

Verify that the binary counter in Fig. 4.8 will count down if either Q is propagated between flip-flops and $\overline{Q}$ used for output, or Q is output and $\overline{Q}$ is propagated. What happens if $\overline{Q}$ is both output and propagated?

## The synchronous trigger flip-flop

A synchronous TFF can be constructed from a master–slave JKFF, by connecting J and K together to form the trigger (T) data input. The clock input receives the control information which determines when the device switches.

The diagram of a JKFF wired up as a synchronous trigger is shown in Fig. 4.9. As J and K are connected together, the input values that can be applied to the internal JKFF are restricted. The truth table for the synchronous TFF will comprise those entries from the JK table where J and K have the same value.

The transition table for the synchronous trigger is

| T | $Q_t \to Q_{t+1}$ |
|---|---|
| 0 | $0 \to 0$ |
| 1 | $0 \to 1$ |
| 0 | $1 \to 1$ |
| 1 | $1 \to 0$ |

To change the output state $(0 \to 1)$ or $(1 \to 0)$ the T input must be 1. If the output is to remain constant $(1 \to 1)$ or $(0 \to 0)$ then T has to be 0. The transition $Q_t \to Q_{t+1}$ takes place when a falling edge occurs on the clock input.

A synchronous TFF cannot be used to count a sequence of pulses applied to its T input. Although the output behaviour is determined by the value of the data at T, the timing of the switching is controlled by the clock. Consider the waveforms T and clock applied to a synchronous trigger:

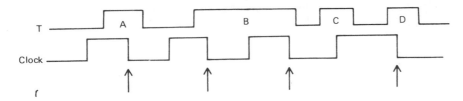

Although the data is continuous, the flip-flop only 'sees' values of T on the falling edges of the clock (indicated by arrows). The flip-flop will therefore completely miss input pulse C and respond twice to pulse B.

Fig. 4.9   A synchronous TFF.

Now consider a clock frequency that is much faster than the data input:

The flip-flop will not change state before time $t_1$ as the input T is 0 on each clock falling edge. At $t_1$ the data input T goes high and the flip-flop output will change state on every falling clock edge, between times $t_1$ and $t_2$. After $t_2$ the input becomes 0 and the output remains constant.

When T = 1 and $Q_t$ = 0, $Q_{t+1}$ = 1, $Q_{t+2}$ = 0, etc.

If the frequency of the clock is known, the number of switching operations in the flip-flop will depend on the duration of the input pulse. Synchronous TFFs can therefore be used in the building block of timing circuits. These will be examined in detail in Chapter 5.

### The delay flip-flop

The delay flip-flop (DFF) is another restricted JKFF. It is sychronous and has a single data input D that is applied directly to J but inverted before going to K. Hence

$$J = D \quad \text{and} \quad K = \overline{D}$$

The block diagram of a DFF, based on a JKFF, is shown in Fig. 4.10.
The transition table can be obtained from the JK truth table. Valid D inputs are when $J = \overline{K}$.

| J | K | $Q_t$ | $Q_{t+1}$ | |
|---|---|---|---|---|
| 0 | 0 | 0 | 0 | |
| 0 | 0 | 1 | 1 | |
| 0 | 1 | 0 | 0 | Valid D = 0 |
| 0 | 1 | 1 | 0 | Valid D = 0 |
| 1 | 0 | 0 | 1 | Valid D = 1 |
| 1 | 0 | 1 | 1 | Valid D = 1 |
| 1 | 1 | 0 | 1 | |
| 1 | 1 | 1 | 0 | |

| D | $Q_t \rightarrow Q_{t+1}$ |
|---|---|
| 0 | $0 \rightarrow 0$ |
| 1 | $0 \rightarrow 1$ |
| 1 | $1 \rightarrow 1$ |
| 0 | $1 \rightarrow 0$ |

Fig. 4.10   A delay flip-flop.

The device is called a DFF because the next output $Q_{t+1}$ always takes the value of the input D, regardless of the present state $Q_t$. It may be used to introduce a controlled delay into a logic system, the delay being dependent on the external clock frequency. If the clock is stopped the DFF will store its last input.

### Shift registers

A shift register may be formed by connecting DFFs in series. The clocks of the individual DFFs are all connected together as shown in Fig. 4.11. After every clock pulse the data stored in each flip-flop is shifted into its neighbour. Data may be input 1 bit at a time to $DFF_0$ and propagated through the shift register to the final output from $DFF_n$. Such devices are known as serial-in, serial-out (SISO) shift registers and can have a capacity of up to 2k-bits.

Compare the shift register with the counter which is asynchronous TFFs in series.

Fig. 4.11   A shift register.

If preset and clear controls are used on each DFF, data can be loaded into the shift register in parallel. Connections from each individual flip-flop output will create a parallel data output path, giving a parallel-in parallel-out (PIPO) shift register. Owing to the large number of input and output lines required by PIPOs, they are restricted to at most 16 bits, in their single-chip integrated form.

### *Shift register applications*

*Parallel-to-serial converter.* Parallel data occurring on several lines may be converted into a string of serial information on a single line by using a parallel-in, serial-out shift register. The data are loaded into the shift register via the parallel inputs. The register is clocked $n$ times, where $n$ is the number of bits in the word. Each clock pulse will cause 1 bit to be output on the serial output line $Z_n$.

'Clocked' means one clock pulse has been applied.

Use a shift register to convert a 4-bit parallel word 1011 into a serial stream of data.

**Worked Example 4.4**

*Solution* A 4-bit parallel-in, serial-out shift register is needed. The word 1011 is applied to the parallel inputs

The states and outputs $Z_n$ are as follows:

| Clock pulse | State | $Z_n$ |
|---|---|---|
| 1 | 1011 | 1 |
| 2 | 0101 | 1 |
| 3 | 0010 | 0 |
| 4 | 0001 | 1 |

The data on $Z_n$ is therefore

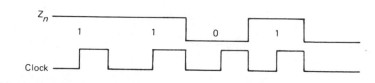

Note. The clock information is needed in order to decode the serial data.

The connections between the data lines and the shift register will determine whether the least or most significant bit is output first on $Z_n$.

One common application of a parallel-to-series converter is in a multiplier circuit where the shift register controls the successive addition.

*Serial-to-parallel converter.* By using the serial input and the parallel outputs of a shift register, a data stream on a serial line can be distributed over the parallel outputs. This operation is the dual of the parallel-to-serial system.

*Shift registers with feedback.* If feedback is applied to a shift register, a repeatable sequence of states may be generated. The serial output or its inverse may be fed back to the serial input or alternatively the input could be a function of two or more of the parallel outputs.

By feeding back the serial output to the input (Fig. 4.12), a shift register can perform a parallel-to-serial conversion without losing its data. If the register's initial state is 1011 the states after each shift and feedback will be

Shift 0 is the initial state.

| Shift | State |
|---|---|
| 0 | 1011 |
| 1 | 1101 |
| 2 | 1110 |
| 3 | 0111 |
| 4 | 1011 |

After four shifts, all the state bits have been output on the serial port, but the

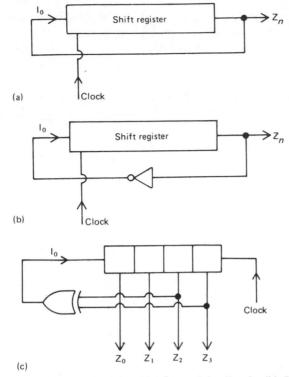

Fig. 4.12. Feedback shift registers. (a) Normal feedback. (b) Inverted feedback.
(c) EXCLUSIVE OR feedback.

original data is still in the register. In general, the data in an $n$-bit register with normal feedback is restored after $n$ shifts.

An inverted feedback shift register has the inverse of its output fed back (Fig. 4.12b).

If an $n$-bit word in an inverted feedback shift register is shifted $n$ times, the original word is inverted. A further $n$ shifts restores the original word.

---

Determine the state sequence of an inverted feedback shift register whose initial state $Q_3Q_2Q_1Q_0$ is 1011 and $I_0 = \overline{Q}_3$.    **Worked Example 4.5**

*Solution*

|               | $Q_3$ | $Q_2$ | $Q_1$ | $Q_0$ |                                                                                 |
|---------------|-------|-------|-------|-------|---------------------------------------------------------------------------------|
| Initial state | 1     | 0     | 1     | 1     |                                                                                 |
| 1st shift     | 0     | 1     | 1     | 0     | $Q_0 \to Q_1, Q_1 \to Q_2, Q_2 \to Q_3$ but $\overline{Q}_3 \to Q_0$            |
| 2nd shift     | 1     | 1     | 0     | 1     |                                                                                 |
| 3rd shift     | 1     | 0     | 1     | 0     |                                                                                 |
| 4th shift     | 0     | 1     | 0     | 0     | Inverse of initial state                                                        |
| 5th shift     | 1     | 0     | 0     | 1     |                                                                                 |
| 6th shift     | 0     | 0     | 1     | 0     |                                                                                 |
| 7th shift     | 0     | 1     | 0     | 1     |                                                                                 |
| 8th shift     | 1     | 0     | 1     | 1     | Original data restored after $2n$ shifts                                        |

If the feedback function is the EXCLUSIVE OR of two parallel output bits as in Fig. 4.12(c) the shift register will generate a sequence of apparently random numbers. The sequence can be repeated provided the shift register always starts in the same state and the feedback function is not changed. The random number sequence contains a maximum of $2^n - 1$ states, where $n$ is the number of bits in the register. Linear feedback shift registers can be used in the coding of information for security purposes, when for example confidential data is transmitted over a public communications channel. The data is encoded with a random number generator and may be decoded if the feedback function and the starting state of the shift register are known.

**Worked Example 4.6**

The message 'This is secret' is stored letter by letter in addresses 1 to 15 of a memory. Devise a feedback shift register to encode the message before transmission and to decode it at the receiver.

*Solution*

Message in transmitter

| Memory address | Contents | Memory address | Contents |
|---|---|---|---|
| 1 | T | 9 | S |
| 2 | H | 10 | E |
| 3 | I | 11 | C |
| 4 | S | 12 | R |
| 5 |   | 13 | E |
| 6 | I | 14 | T |
| 7 | S | 15 | . |
| 8 |   |   |   |

FSR is 'feedback shift register'. Shift register stores word $Q_3Q_2Q_1Q_0$ and $I_0$ is its serial input.

Now consider a 4-bit FSR with $I_0 = Q_2 \oplus Q_3$ and starting state 1011. Its number sequence is

After each shift $Q_2 \rightarrow Q_3$, $Q_1 \rightarrow Q_2$, $Q_0 \rightarrow Q_1$, but $Q_3 \oplus Q_2 \rightarrow Q_0$.

|  | $Q_3$ | $Q_2$ | $Q_1$ | $Q_0$ | Decimal |
|---|---|---|---|---|---|
| Initial state | 1 | 0 | 1 | 1 | 11 |
| 1st shift | 0 | 1 | 1 | 1 | 7 |
|  | 1 | 1 | 1 | 1 | 15 |
|  | 1 | 1 | 1 | 0 | 14 |
|  | 1 | 1 | 0 | 0 | 12 |
|  | 1 | 0 | 0 | 0 | 8 |
|  | 0 | 0 | 0 | 1 | 1 |
|  | 0 | 0 | 1 | 0 | 2 |
|  | 0 | 1 | 0 | 0 | 4 |
|  | 1 | 0 | 0 | 1 | 9 |
|  | 0 | 0 | 1 | 1 | 3 |
|  | 0 | 1 | 1 | 0 | 6 |
|  | 1 | 1 | 0 | 1 | 13 |
|  | 1 | 0 | 1 | 0 | 10 |
|  | 0 | 1 | 0 | 1 | 5 |
| (Repeat sequence) | 1 | 0 | 1 | 1 | 11 |

82

The random number sequence is used to address the transmitter memory and the message transmitted will be

CS.TR THSSIIEE

At the receiver the message can be decoded by using the same random sequence. The complete set of numbers DOES NOT have to be transmitted. It can be generated in the receiver with another FSR, provided both the feedback function $(I_0 = Q_3 \oplus Q_2)$ and the initial state 1011 are known. The letters received will be written into a memory whose addresses are determined by the FSR.

| Letter received | Address | Letter received | Address |
|---|---|---|---|
| C | 11 | S | 4 |
| S | 7 | S | 9 |
| . | 15 | I | 3 |
| T | 14 | I | 6 |
| R | 12 | E | 13 |
|  | 8 | E | 10 |
| T | 1 |  | 5 |
| H | 2 |  |  |

The decoded message in the receiver memory will be

| Memory address | Contents | Memory address | Contents |
|---|---|---|---|
| 1 | T | 9 | S |
| 2 | H | 10 | E |
| 3 | I | 11 | C |
| 4 | S | 12 | R |
| 5 |  | 13 | E |
| 6 | I | 14 | T |
| 7 | S | 15 | . |
| 8 |  |  |  |

## Summary

A sequential circuit is formed by applying feedback to a combinational logic system. Feedback creates additional internal inputs (or states) that cannot be controlled directly by the external inputs. They are dependent on the previous inputs and state. The latter is dependent on earlier inputs and states, and so on. Hence a sequential circuit's behaviour is a function of all its previous inputs.

The stability of a sequential circuit can be summarized on a flow table. For a given value of external inputs, the circuit will always switch into a stable state unless a recurring cycle of unstable states are entered, causing an oscillation.

The SRFF is a sequential circuit with a single feedback loop. Its circuit is a pair of cross-coupled gates (either NAND or NOR). It is a memory device, but has inherent timing problems. The timing problems in the SRFF are overcome in the master–slave JKFF.

If J and K are permanently set to 1 and data entered via the clock input, we have the asynchronous trigger which forms the basis of many counting circuits. With J and K connected together to form a common data input and a separate clock, we have the synchronous trigger that can be used in timing circuits. Finally, if an inverter is then inserted between J and K such that the data input goes to J and its inverse to K, the result is a DFF.

DFFs connected in series form a shift register which has wide applications when data has to be converted between parallel and serial forms. A shift register with feedback can be used to retain information, to invert binary words and to create sequences of random numbers.

## Problems

4.1 Analyse the following sequential circuit and summarize its behaviour on a flow table:

$$F = A + \overline{\overline{B}.\overline{F'}}$$

where F is fed back to form internal input F'.

4.2 An SRFF is driven from a 2-bit binary counter $b_1 b_0$. Output $b_1$ is ANDed with $b_1 \oplus b_0$ and drives R. $b_0$ is also ANDed with $b_1 \oplus b_0$ and drives S. Evaluate the output from the SRFF as the counter progresses through its four states starting at 00. (The flip-flop is initially reset to 0.)

4.3 A JKFF is driven from a 2-bit binary counter with $b_1$ connected to J and $b_0$ connected to K. The flip-flop and counter have a common clock and the flip-flop is initially reset to 0. What is its output sequence as the counter is incremented from 00? Comment on any possible timing problems.

4.4 Design a master–slave JKFF using only NAND gates. What is the minimum number of gates required?

4.5 A sequential system comprises two feedback logic functions

$$F_1 = AB\overline{F'_1} + F'_1\overline{F'_2}$$
$$F_2 = (A + B + F'_2)(\overline{A} + \overline{B} + F'_2)$$

where $F_1 \to F'_1$ and $F_2 \to F'_2$. Under what input conditions will the system oscillate?

4.6 A manufacturing error results in a batch of flip-flops with cross-coupled NOR/NAND gates as shown below:

Show that this device *cannot* be reset via its input terminals AB.

4.7 A circuit comprises a synchronous trigger TFF0, driven by input I. Its output $Q_0$ drives a delay flip-flop DFF1 whose output, $Q_1$, in turn drives another synchronous trigger TFF2. Determine the sequence of outputs on $Q_2Q_1Q_0$, starting from 000, that occur when I is set to 1 and the circuit clocked.

4.8 Design a logic system using two SRFFs and any necessary AND/OR/NOT logic to control two machines. Each machine is turned on by a flip-flop if its output is 1 and off if 0. The logic system operates from three push buttons A, B and C. When A is pressed, both machines are switched on. If both machines are on and B is pressed, machine 1 must be turned off. Any machine that is on when C is pressed, is turned off. Also, devise a digital input circuit to prevent the input data from being transferred to the controller if two or more buttons are pressed simultaneously.

4.9 Determine the count sequence, starting at 000, when three synchronous triggers are connected in series, the input set to 1 and the system clocked.

Compare and contrast this counter with a 3-bit version of the asynchronous counter given in Fig. 4.8.

4.10 If TFF2 in the counter of Problem 4.9 is replaced with a data flip-flop, what will the new count sequence be?

4.11 A 3-bit register $Q_2Q_1Q_0$ has the feedback function $I_0 = Q_2 \oplus Q_1$. What sequence does it generate if the original state is 001?

4.12 A 4-bit shift register $Q_3Q_2Q_1Q_0$ has a feedback function

$$I_0 = \overline{Q_3 \cdot (Q_2 \oplus Q_1)}$$

where $I_0$ is the serial input. What is its output sequence if its starting state is 0000?

# 5 Design of sequential logic circuits

**Objectives**
- ☐ To identify the parameters and structure of a general sequential circuit.
- ☐ To design asynchronous counting circuits using trigger flip-flops and to investigate the use of output encoders to enable count sequences other than pure binary to be implemented.
- ☐ To examine the limitations of asynchronous circuits and the use of synchronous circuits to overcome these practical problems.
- ☐ To evaluate the JKFF as a building block for synchronous systems and the design of its driving logic.
- ☐ To specify sequential systems by means of present/next state tables.
- ☐ To apply state transition diagrams to sequential logic design.
- ☐ To simplify sequential systems by state minimization.
- ☐ To develop a tabular method for state minimization.
- ☐ To develop design strategies for asynchronous sequential logic.
- ☐ To introduce algorithmic state machine charts.

In Chapter 4, we saw that a sequential circuit was basically a combinational logic circuit with feedback, and a range of simple feedback circuits called flip-flops could be devised. This chapter will concentrate on the design of logic systems using these discrete building blocks. As every sequential circuit contains a large amount of combinational logic, the methods and techniques outlined in Chapter 3 will be required when designing sequential systems.

### The parameters of a sequential circuit

The internal state is often simply referred to as the 'state'. $\delta$ is pronounced 'delta'.

Note. The output of $\delta$ (the next state) when fed back becomes the internal inputs (the new present state).

$\omega$ is the Greek letter omega.

In any sequential logic system three distinct sets of data and two logic functions can be identified. The input data is represented by the set **I** and the output by **Z**. The sequential system generates internal data or states, which form additional internal inputs to the logic functions. These internal states form the data set **Q**. The logic function that generates the internal states has the symbol $\delta$, and this is implemented by a combinational logic circuit operating on the external inputs and the internal state (the outputs of $\delta$) fed back to form internal inputs. The output of $\delta$ is the next state of the system and the data on the internal inputs is the present state.

The next state logic system is responsible for the circuit's sequential properties that arise from the feedback connections. This part of the circuit can be partitioned into flip-flops – the standard circuits with a single feedback connection which were examined in the previous chapter.

The second logic function within a general sequential system is $\omega$, a combinational logic circuit which generates the output **Z**. Its inputs are the external inputs **I** and the state **Q**. The general form of a sequential logic system is given in Fig. 5.1.

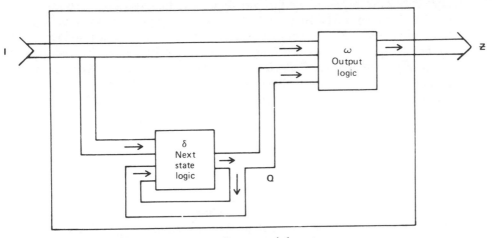

Fig. 5.1   A general sequential system.

The output logic can be obtained from the truth table which specifies the outputs for all possible combinations of inputs and internal states. In some sequential circuits there is no need for output logic, as the outputs are identical to the internal states. Counters and code generators do not require any output logic, unlike 'one in $n$' and sequence detectors where the output is a function of the inputs and the state of the system.

See Worked Example 5.6.

The feedback applied to the next state logic is crucial to sequential behaviour. The logic outputs that represent the internal state of the system are fed back to provide internal inputs. This enables the next state logic to give different responses to identical external inputs applied to the circuit at different times in a sequence of inputs. Suppose a logic system has two inputs $I_1$ and $I_2$, and two state variables $Q_1$ and $Q_2$. Let the external inputs be 00 and the state also 00. The four inputs to the next state logic will be $I_1I_2Q_1Q_2 = 0000$ and an output will be generated. Let the new output, which is the next state of the system, be 01. This is fed back and the input data to the next state logic becomes 0001. The logic designer can ensure that a new output is generated from this data. Suppose the new output is 11. It is now possible to distinguish between the first and second occurrences of 00 on the inputs. When 00 is first applied, the internal state becomes 01. A subsequent input of 00 switches the state to 11. Detection of the different occurrences of input data, in a sequence, is possible even though the data is numerically equal to earlier or later inputs. Hence the system is sensitive to the sequence of the inputs.

This would only be attempted in a synchronous system where the inputs are synchronized with a clock.

The designer of the next state logic has to be able to identify the internal states of the system and then design hardware to generate them. The hardware can be implemented in one of three forms. It can be based on

(i) combinational logic with feedback or
(ii) standard flip-flops or
(iii) semiconductor memory.

Sequential system comprising combinational logic with feedback will not be examined in great depth here. Although these circuits can operate at very high

Feedback combinational circuits are sometimes called primitive circuits.

speeds, they are prone to timing problems. The logic is asynchronous and any unequal delays in the feedback paths can cause the system to malfunction.

The use of VLSI memory devices as building blocks for sequential logic systems will be examined in Chapter 6. These devices are suitable for larger systems and represent an extension of the principles pertaining to designs based on standard flip-flops which are, in effect, 1-bit memories. The remainder of this chapter will be devoted to the design of sequential logic using discrete flip-flops.

VLSI: Very Large Scale Integration.

### Asynchronous binary and non-binary counters

Take care not to confuse 'asynchronous circuit' with 'a synchronous circuit' – they sound the same.

In Chapter 4 it was found that an asynchronous TFF would divide an input pulse train by 2 so that the output contained only half the number of pulses in the input. A binary counter can be formed by cascading several TFFs as shown in Fig. 5.2 This is, in effect, $\delta$, the next state logic for the counter.

Fig. 5.2  An asynchronous binary counter.

Asynchronous TFFs switch on the falling edge of the inputs. The counter will be insensitive to the length of time the input is held at 1.

A counter is a sequential logic circuit. A differential output $(Q_2Q_1Q_0)$ equal to the value of the count is required each time the input I is set to 1. The circuit in Fig. 5.2 does not depart from the general concept of the feedback circuit given in Fig. 5.1. The external input together with the internal state variables drives the next state logic which consists of three TFFs. The next state logic can be redrawn as a feedback circuit as shown in Fig. 5.3.

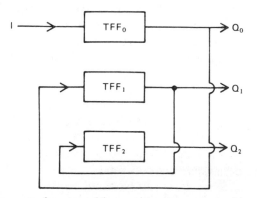

Fig. 5.3  An asynchronous binary counter arranged as a feedback circuit.

The 'driving logic' is sometimes called the 'excitation' or 'steering logic'.

This is a very simple sequential circuit. There is no output logic and the feedback is connected directly to the terminals of the flip-flops. In more complicated systems the inputs to the flip-flops will be functions of both input and internal state variables. These functions are known as the 'driving logic'. A general block diagram of next state logic with driving functions is shown in Fig. 5.4.

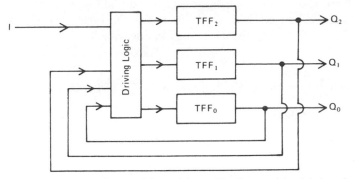

Fig. 5.4 Next state logic partitioned into flip-flops with driving functions.

If a non-binary count is required, driving logic is needed to control the flip-flops, and its design is illustrated in the following worked example.

**Worked Example 5.1**

Design an asynchronous counter that, on the falling edge of each input pulse, progresses through the sequence 000, 010, 001, 011, 100, and repeats from 000.

*Solution*  If asynchronous TFFs are used, the count sequence must be examined in order to identify falling edges which can then be used to switch the flip-flops.

Each state variable change must be identified, whether 0 to 1 or 1 to 0.

| $Q_2$ | $Q_1$ | $Q_0$ | State |
|-------|-------|-------|-------|
| 0 | 0* | 0 | a |
| 0 | 1* | 0* | b |
| 0 | 0* | 1 | c |
| 0* | 1* | 1* | d |
| 1* | 0 | 0 | e |
| 0 | 0 | 0 | a (repeated) |

Each change of state of flip-flop output is denoted with an asterisk. It can be seen that the input I cannot be used to switch any of the flip-flops, as none of the output bits changes state on every input pulse. $Q_0$ changes its value as the count

A falling edge must occur on the input I before every change in the count.

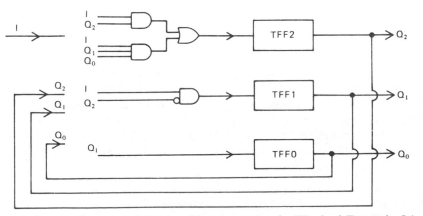

Fig. 5.5 Logic diagram of the non-binary counter in Worked Example 5.1.

progresses from state b to c and from d to e. The falling edge on $Q_1$ occurring after state b and d can be used to switch TFF0. On further examination of the state sequence the following functions may be identified:

$$T_1 = I\overline{Q_2}$$

I must be included in the driving functions to prevent the counter from halting in states a or e.

which will switch the second flip-flop giving bit $Q_1$ of the count and

$$T_2 = I(Q_1 Q_0 + Q_2)$$

which will switch the third flip-flop producing the most significant bit of the count.

The complete logic diagram for the system is given in Fig. 5.5.

**Exercise 5.1**

Design an asynchronous counter using TFFs that will count in BCD. The counter must comprise 4 bits and the count sequence, starting at 0000, proceeds to 1001 and then returns to 0000 and repeats the sequence.

*Practical limitations of asynchronous counters*

The design procedure when using edge-triggered flip-flops involves the identification of functions that will produce falling edges at points in the count sequence which coincide with the changing of one or more bits in the count. The edges can then be used to switch the flip-flops.

Two problems arise from this approach:

(i) It may not be possible to identify functions to give the desired flip-flop switching sequence.
(ii) The count may halt in certain states if the switching of two or more bits are inter-dependent.

Problem (ii) can be illustrated as follows. Suppose a count sequence required a 2-bit counter $Q_1 Q_0$ to change from 11 to 00. The falling edge on $Q_1$ appears to be suitable to switch flip-flop $T_0$ and likewise $T_1 = Q_0$. In practice, however, each flip-flop would be waiting for the other to change and if there was no other data applied to the trigger inputs, the counter would halt in state 11.

As a result of these limitations, it is not possible to design certain non-binary sequence counters using asynchronous triggers and driving logic.

*An asynchronous counter with output logic*

The general form of a sequential circuit (Fig. 4.1) contains two logic systems, $\delta$, the next state logic, and $\omega$, the output logic. However, the counters in the previous section of this chapter did not require any output logic as the outputs were identical to the internal states. If a particular count sequence cannot be achieved using only next state logic because of the problems of identifying suitable switching functions and their possible inter-dependence, a pure binary counter with output logic has to be used. The output logic then encodes the internal states into the desired count sequence.

**Worked Example 5.2**

Design a 3-bit counter that will count a string of randomly occurring pulses in Gray code.

*Solution* The pulses can be counted in pure binary by cascading three TFFs (see Fig. 5.2). Output logic is required to convert the binary to Gray code according to the following truth table:

| $Q_2$ | $Q_1$ | $Q_0$ | $Z_2$ | $Z_1$ | $Z_0$ |
|---|---|---|---|---|---|
| 0 | 0 | 0 | 0 | 0 | 0 |
| 0 | 0 | 1 | 0 | 0 | 1 |
| 0 | 1 | 0 | 0 | 1 | 1 |
| 0 | 1 | 1 | 0 | 1 | 0 |
| 1 | 0 | 0 | 1 | 1 | 0 |
| 1 | 0 | 1 | 1 | 1 | 1 |
| 1 | 1 | 0 | 1 | 0 | 1 |
| 1 | 1 | 1 | 1 | 0 | 0 |

$Z_2$, $Z_1$ and $Z_0$, are all functions of $Q_2Q_1Q_0$ and can be minimized on K-maps.

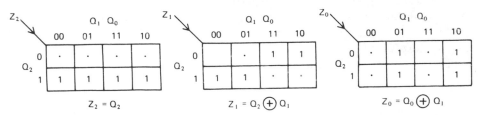

The EXCLUSIVE OR function is defined as $Q_2 \oplus Q_1 = Q_2\overline{Q_1} + \overline{Q_2}Q_1$. $Z_2 = Q_2$ can be observed on the truth table.

The complete logic diagram for the system is given in Fig. 5.6.

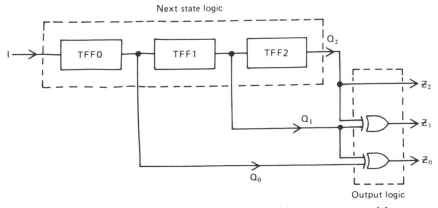

Fig. 5.6 A Gray code counter based on a pure binary counter with output logic.

A counter can be designed to progress through any code, if output logic is included in the sequential system. There is, however, one shortcoming that must be taken into account in practical circuits. The next state information may not be output from the logic simultaneously. For example, a longer time is required to generate $Q_2$ than for $Q_1$ or $Q_0$ in the pure binary counter because of the different

The effects of next state propagation delays are eliminated in a synchronous system.

Hazards arise from unequal propagation paths in combinational logic. A delay of 5 nanoseconds to 1 microsecond, depending on the type of logic being used, is required for the hazards to pass.

propagation paths. These unequal delays, together with hazards in the output logic, mean that a finite time must elaspse after a pulse has been input, before the updated value of the count on the output is valid.

**Exercise 5.2**

Design an asynchronous counter to increment in the following sequence: 000,010,011,100, 111,101,001,110,000 etc.

(Investigate whether the counter can be built using next state logic only. If not, use a binary counter with output logic.)

### Synchronous sequential logic

In a synchronous sequential logic system the switching of the circuit depends solely on the input data, but the timing within the circuit is controlled by an external clock. The clock consists of a stream of pulses, and each pulse allows the logic to switch once only, and this allows us to be more precise about the definitions of the present and next states of the system.

In an asynchronous system the present state is defined by the internal inputs $Q$ at the instant external data is applied. The next state is the value of $Q$ when the logic has responded to the inputs. The time taken to generate the next state depends solely on the characteristics of the logic circuits and is normally no more than a few microseconds. The next state is fed back and becomes the new present state and the circuit will operate again. In asynchronous circuits it is highly probable that the value of the internal state will be corrupted as it is fed back. This is due to unequal propagation delays in the logic.

In asynchronous circuits it is not possible to distinguish between consecutive present states or inputs which are numerically equal.

In synchronous sequential logic the present state of the system is its internal state before a clock pulse has been applied. When a clock pulse occurs, the system switches. The next state is generated and is fed back to the inputs of the logic to become the new present state. However, no further switching occurs until the next clock pulse arrives. During this time errors due to propagation delays will disappear and internal states can be precisely identified and generated at the rate of one state per clock pulse.

The period of the clock waveform must be greater than any propagation delays in the system.

*Present/next state table*

A present/next state table relates the next state of a sequential system to the external inputs and the present state. The structure of the present/next state table is given in Table 5.1.

**Table 5.1   A present/next state table**

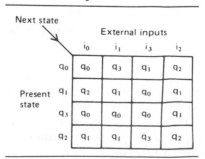

| Next state → External inputs | $i_0$ | $i_1$ | $i_3$ | $i_2$ |
|---|---|---|---|---|
| Present state $q_0$ | $q_0$ | $q_3$ | $q_1$ | $q_2$ |
| $q_1$ | $q_2$ | $q_1$ | $q_0$ | $q_1$ |
| $q_3$ | $q_0$ | $q_0$ | $q_0$ | $q_1$ |
| $q_2$ | $q_1$ | $q_1$ | $q_3$ | $q_2$ |

The columns of the table represent the value of the inputs and the rows are labelled with the present state of the system. The contents of the cells give the value of the next state after a clock pulse has been applied. State sequences can therefore be identified on a present/next state table.

Unlike the K-map, the present/next state table can have unlimited rows and columns.

---

**Worked Example 5.3**

A 3-bit up/down counter counts up if its external input I is 1 and counts down if I is 0. It has an upper limit of $7_{10}$ and a lower limit of $0_{10}$. Obtain a present/next state table for this counter.

*Solution*  Let the eight states be $q_0$ to $q_7$, corresponding to $000_2$ to $111_2$ ($0_{10}$ to $7_{10}$). If I = 1,

then  $q_{t+1} \rightarrow q_t + 1$  for $0 \leqslant t < 7$
else  $q_{t+1} \rightarrow q_t$      for $t = 7$

Similarly, if I = 0

then  $q_{t+1} \rightarrow q_t - 1$  for $0 < t \leqslant 7$
else  $q_{t+1} \rightarrow q_t$      for $t = 0$

The present/next state table that describes this behaviour is therefore

| Present | Next | |
|---------|------|------|
|         | I = 1 | I = 0 |
| $q_0$ | $q_0$ | $q_1$ |
| $q_1$ | $q_0$ | $q_2$ |
| $q_2$ | $q_1$ | $q_3$ |
| $q_3$ | $q_2$ | $q_4$ |
| $q_4$ | $q_3$ | $q_5$ |
| $q_5$ | $q_4$ | $q_6$ |
| $q_6$ | $q_5$ | $q_7$ |
| $q_7$ | $q_6$ | $q_7$ |

---

*Design of synchronous next state logic*

The next state logic in a synchronous sequential system can be based on either JKFFs or synchronous TFFs. The output of each device contributes 1 bit to the value of the internal state and the switching pattern for each flip-flop can be obtained from the present/next state table. Transition tables for both the JKFF and the synchronous TFF were developed in Chapter 4. By referring to the appropriate transition table for the particular type of flip-flop to be used in a system, the logic values required on the outputs of the flip-flops may be identified, and then the driving logic function obtained. The complete design process will be illustrated in the following worked example.

---

**Worked Example 5.4**

Design a timer having 3-bits resolution that measures the duration of an input pulse.

*Solution*  A block diagram of the system has a single external input I and three outputs to display the value of the count to 3-bits accuracy.

An external clock whose frequency must be known is also required.

A present/next state table for the timer can be compiled. There are two input states as the single input variable I can be either 0 or 1 and eight internal states giving the value of the count from 000 to 111. No output logic is required in this system. The outputs are identical to the internal states. The present/next state table will therefore require two columns, $I = 0$ and $I = 1$, and eight rows, $Q_2Q_1Q_0 = 000$ to 111. The contents of each cell will be the next value of $Q_2Q_1Q_0$ following a clock pulse.

Cell $(I = 0, Q_2Q_1Q_0 = 000)$ represents the state of the timer before $t_1$. The clock is running but the input pulse has not yet occurred. The counter must therefore remain in state 000. When the input becomes 1 the next state of the counter at each clock pulse is the present state plus 1. For non-zero values of the count and $I = 0$, the next state of the counter needs to be the same as the present state. This condition applies after time $t_2$ where the input has passed and the duration of the pulse needs to be held in the system. The complete present/next state table for the timer is:

The unit of time in the system is the clock period.

| Next state | | $I = 0$ | $I = 1$ |
|---|---|---|---|
| Present | 000 | 000 | 001 |
| state | 001 | 001 | 010 |
| $Q_2Q_1Q_0$ | 010 | 010 | 011 |
| | 011 | 011 | 100 |
| | 100 | 100 | 101 |
| | 101 | 101 | 110 |
| | 110 | 110 | 111 |
| | 111 | 111 | 000 |

The designer can now establish how many flip-flops are required in the system. In general, $n$ flip-flops can represent $2^n$ states and $n$ must be an integer. Therefore given the total number of internal states $x$, $n$ is the integer power of 2 equal to $x$ or the first integer power of 2 greater than $x$. In this design $x$ is 8 and $2^3 = 8$. Therefore $n$, the number of flip-flops required, is 3.

Either JKFFs or synchronous TFFs may be used in the next state logic. In this example we shall use synchronous triggers and the reader can repeat the design for him- or herself using JKFFs. The output of each synchronous trigger contributes 1 bit to the value of the state. The present/next state table specifies the state transitions within the system for all possible values of inputs. For example, with the present state 000 and $I = 0$, the next state is 000. The flip-flop representing the least significant bit of the state remains at 0. Its present to next state transition is therefore 0 to 0. However, for the same state but external input $I = 1$, the least significant bit must switch from 0 to 1. The output transitions of the flip-flops for all possible input conditions can therefore be obtained from the present/next state table. The inputs to the flip-flops now have to be identified to give these desired output switchings. The transition table for the synchronous TFF is

A TFF requires one driving function (the T input). A JKFF requires two; however, they are usually simpler, owing to the 'don't cares' in the JKFF transition table.

| Input T | Output transition |
|---------|-------------------|
| 0 | $0 \rightarrow 0$ |
| 1 | $0 \rightarrow 1$ |
| 0 | $1 \rightarrow 1$ |
| 1 | $1 \rightarrow 0$ |

The driving logic functions $T_0$, $T_1$ and $T_2$ can now be designed. They are all functions of $I$, $Q_2$, $Q_1$ and $Q_0$. A K-map may be used if the functions are required in minimal form. The map for $T_2$ is

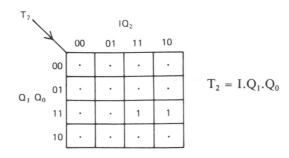

$$T_2 = I.Q_1.Q_0$$

Remember. The axes of the K-maps must be labelled in Gray code.

. on K-map indicates logical 0.

The contents of each cell of the K-map can be deduced from the system's present/next state table and flip-flop transition table. For example cell $IQ_2 = 00$, $Q_2Q_1 = 00$ on the K-map corresponds to state $I = 0$, $Q_2Q_1Q_0 = 000$ on the present/next state table. The K-map represents the function $T_2$, which is driving the flip-flop producing the most significant bit $Q_2$ of the internal state. From the present/next state table we see that the present value of $Q_2$ is 0 and its next value is also 0. From the transition table for the TFF, 0 to 0 on the output requires T to be 0 and this is entered into the K-map. The procedure is repeated for the 15 remaining input conditions and the K-map completed. The driving function $T_2$

may then be obtained from the K-map. It is

$$T_2 = IQ_1Q_0$$

K-maps for $T_1$ which produces bit 1 of the internal state and $T_0$, which generates the least significant state bit $Q_0$, can also be plotted. They are

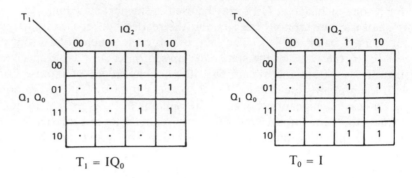

$$T_1 = IQ_0 \qquad\qquad T_0 = I$$

The logic diagram for the complete timer is given in Fig. 5.7.

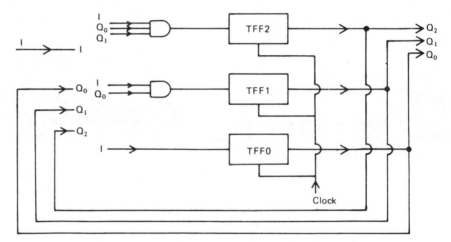

Fig. 5.7   Logic diagram for a 3-bit timer.

The timer will be accurate within one clock pulse regardless of the value of the count. The accuracy will also depend on the precision of the clock frequency. In this circuit it is essential to have an external clear facility to each of the flip-flops in order to reset the timer to zero after use.

**Exercise 5.3**   Redesign the 3-bit timer in the above worked example using JKFFs in the next state logic.

**Cellular sequential logic**

A formal design of a system having many variables would be very time consuming.

In combinational logic design we saw that cellular or iterative circuit forms are very desirable as they enable the resolution in the logic system to be increased without requiring a total redesign of the system. Additional standard modules are simply added to the existing hardware. Cellular circuit forms are most useful when fabricating large systems in VLSI as the cell can be repeated automatically on the mask. The design process is reduced to the identification of a suitable cellular structure and the logical design of the cell.

These advantages are equally valid for cellular sequential circuits. In the timer circuit (see Worked Example 5.4) the driving logic functions were

$$T_0 = I$$
$$T_1 = IQ_0$$
$$T_2 = IQ_0Q_1$$

A regular pattern emerges from these functions. The general function $T_n$ – the trigger function producing the $n$th significant bit of the count – is the AND of the input I and the present values of all the less significant bits:

$$T_n = IQ_{n-1}Q_{n-2}\cdots Q_2Q_1Q_0$$

Hence the system could be expanded to $n$-bits resolution. The $n$th flip-flop would, however, require a driving logic function comprising an AND gate with $n$ inputs. This function would become unwieldy for large values of $n$.

Comparing $T_2$ with $T_1$ we see that

$$T_2 = IQ_0Q_1 = T_1Q_1$$

because

$$T_1 = IQ_0$$

Therefore

$$T_n = Q_{n-1}T_{n-1}$$

For any flip-flop in the timer, the driving function is the AND of the output of the previous stage $Q_{n-1}$ and the driving function to that stage. The circuit can therefore be built out of these standard cells and the resolution depends solely on the number of cells in the system as shown in Fig. 5.8.

Fig. 5.8   An iterative cellular version of a timer.

Fig. 5.9   An autonomous sequential circuit.

### Autonomous sequential circuits

An autonomous sequential circuit does not have any external data inputs. The output logic and next state logic are functions of the internal states only. A synchronous system will, of course, require a controlling clock. The block diagram of a general autonomous sequential circuit is given in Fig. 5.9.

Code generators are a simple form of autonomous circuit and their design will be illustrated in the following worked example.

**Worked Example 5.5**   Design a code generator based on JKFFs that produces the following code cycle at the rate of one word per clock pulse:

$$000 \rightarrow 001 \rightarrow 011 \rightarrow 110 \rightarrow 100$$

*The logic is autonomous as there is no external input to influence the next state.*

*Solution*   Three flip-flops are required to represent the 3-bit codeword. The present/next state table in an autonomous circuit consists of two columns, the present state and the next state as shown below:

| Present state | | | Next state | | |
|---|---|---|---|---|---|
| $Q_2$ | $Q_1$ | $Q_0$ | $Q_2$ | $Q_1$ | $Q_0$ |
| 0 | 0 | 0 | 0 | 0 | 1 |
| 0 | 0 | 1 | 0 | 1 | 1 |
| 0 | 1 | 0 | d | d | d |
| 0 | 1 | 1 | 1 | 1 | 0 |
| 1 | 0 | 0 | 0 | 0 | 0 |
| 1 | 0 | 1 | d | d | d |
| 1 | 1 | 0 | 1 | 0 | 0 |
| 1 | 1 | 1 | d | d | d |

*Remember. d is 'don't care'. The present state does not occur in the desired sequence, so its next state is hypothetical.*

The 'don't care' conditions in the present/next state table are important. They occur because the code sequence contains only five different binary words. This means three flip-flops are required that are capable of representing a total of eight states. Three binary words do not occur in the code sequence. They are 010, 101 and 111. If the system functions correctly these values cannot exist as present states and therefore their corresponding next states may be regarded as 'don't cares', which can be carried through to the driving logic.

The driving functions for the flip-flops can be specified by referring to the present/next state table and the JK transition table. For flip-flop 1 (giving the least significant state bit $Q_0$) K-maps for driving functions $J_1$ and $K_1$ which are both the functions of $Q_2 Q_1$ and $Q_0$ are as follows:

JKFF transition table:

| J | K | $Q_r \rightarrow Q_{r+1}$ |
|---|---|---|
| 0 | d | $0 \rightarrow 0$ |
| 1 | d | $0 \rightarrow 1$ |
| d | 0 | $1 \rightarrow 1$ |
| d | 1 | $1 \rightarrow 0$ |

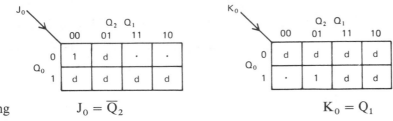

giving $\qquad$ $J_0 = \overline{Q}_2$ $\qquad\qquad\qquad$ $K_0 = Q_1$

By a similar process the following functions can be obtained:

$$J_1 = Q_2 \quad K_1 = \overline{Q}_2$$
$$J_2 = Q_1 \quad K_2 = \overline{Q}_1$$

In this example the driving logic simply consists of connections between the outputs of the flip-flops (the state variables) and their inputs. In more complicated systems the flip-flop inputs may be logical functions of their outputs.

If asynchronous data is applied to the clock, the 'code generator' type of circuit acts as a non-binary counter. This type of circuit does not have a controlling clock independent of the data. The switching is, however, synchronized by the input data. It is a data-synchronized circuit. It is therefore possible to use a synchronous counter with its clock driven by input data as an alternative to asynchronous design with edge-triggered flip-flops. Referring back to Worked Example 5.1, that asynchronous system could equally well have been organized as a synchronous autonomous machine, based on JKFF's or TFFs to generate the desired count sequence. If so, the clock would be driven from the asynchronous data input I.

An example of asynchronous data is a randomly occurring sequence of pulses.

In a simple display unit the binary words 0001 to 1111 represent the letters A to O respectively and 0000 represents a space.

Design a sequential system which will print out the name JOE KING continuously on the VDU.

**Exercise 5.4**

VDU: Visual Display Unit.

### State transition diagrams

In simple sequential systems one can write down the present/next state table straight from the problem specification. In more complex systems where the internal states are not obvious on inspection, the system can be modelled on a state transition diagram and the input states identified. A state transition diagram is a graphical representation of a sequential system. It consists of nodes, representing the internal states, and flow lines between nodes, which are labelled with the inputs causing the state transition and the resulting output. The state transition diagram for the JKFF is given in Fig. 5.10. There are two nodes $q_0$ and $q_1$ representing the internal states 0 and 1. The flow lines are labelled JK/Q where J and K are the inputs and Q is the output.

This form of state transition diagram is known as the Mealy model. Another standard model is due to Moore, and has its nodes labelled with the internal state and the resulting output.

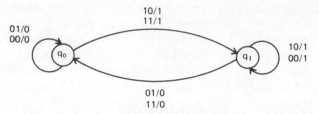

Fig. 5.10   A state transition diagram for a JKFF.

If the flip-flop is in state 0 ($q_0$) and J = 0 and K = 1, the next state of the system is 0. This is represented by the flow line leaving $q_0$ and returning to $q_0$. If, however, J = 1 and K = 0, the flip-flop will be set to 1 and its output goes to 1. This condition is represented by the flow line leaving $q_0$ directed towards $q_1$.

The present/next state table can be obtained from the state transition diagram. Values of the inputs are given on the flow line labels.

Every flow line starts from, and terminates in, a node. The starting node represents the present state and the final node the next state. In Fig. 5.10, if $q_0$ is the present state and the input is 10 the next state will be $q_1$, whereas present state $q_1$ and input 10 results in next state $q_1$.

The present/next state table can be produced by considering all the transitions between nodes on the state transition diagram. For the sequential system in Fig. 5.10 the table is

A flow line may connect different nodes or form a loop leaving and then re-entering the same node.

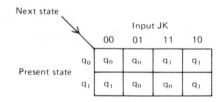

At this stage the states have to be assigned binary values. The next state logic can then be designed using the present/next state table. As we started with a JKFF, the reader should find that the next state logic consists of one flip-flop and the inputs are identical to the variables labelled on the state transition diagram, where $q_0$ is 0 and $q_1$ is 1.

The assignment of binary values to the states is arbitrary, but some state assignments may lead to marginally simpler logic circuits than others.

---

**Worked Example 5.6**

Design a sequential logic circuit that will detect the sequence 101 in a stream of serial data, input at the rate of 1 bit per clock pulse.

*Solution*   The circuit has a single input variable I which is either 0 or 1, and a single output Z that becomes 1 when the input sequence 101 has been received. However, it is not obvious at the outset how many internal states are required. The internal states are identifiable from a state transition diagram.

Let internal state A be the initial state where none of the input bits have arrived in the correct sequence.

In state A, the external input can be either 0 or 1. If I = 0 we remain in state A as the input does not correspond with the first bit in the sequence we wish to

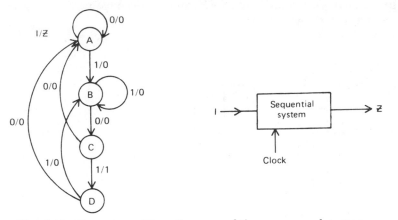

Fig. 5.11   State transition diagram of the sequence detector.

detect. If, however, I = 1 the first bit in the sequence has occurred. This condition is represented by new internal state B. The output in both cases is zero as the complete sequence has not yet been detected. When in state B further input data can be received. If I = 0 the second bit of the sequence 101 has arrived and we switch to state C. If I = 1 the current input might still be the first bit of the sequence 101 and the system should remain in B. The complete state transition diagram is given in Fig. 5.11.

The state diagram contains four nodes and state D is entered after the complete input sequence has been detected. The present/next state table is

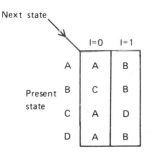

Assigning binary values to the internal states (let A = 00, B = 01, C = 11 and D = 10) we obtain

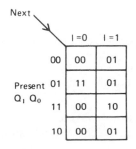

As the assignment of binary values is arbitrary, there are a large number of different hardware solutions to this problem. The state is made up from the outputs of two flip-flops $Q_1 Q_0$.

2$^n$ states require $n$ flip-flops.

The next state logic requires two JKFFs and the driving functions may be derived from K-maps as shown below:

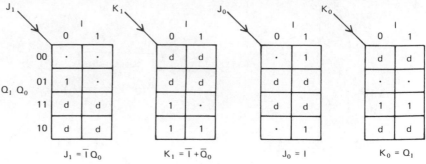

Refer to Worked Example 5.5 for derivation of K-maps from state tables.

$$J_1 = \overline{I}\,Q_0 \qquad K_1 = \overline{I} + \overline{Q}_0 \qquad J_0 = I \qquad K_0 = Q_1$$

See Fig. 5.1.

In this system output logic is required as the output is not identical to the value of the internal states. From the general form of the sequential circuit we see that the output is a function of both the input and the internal state variables.

It is evident from the state transition diagram that the output becomes 1 when in present state C and the input is 1 (the flow line from C to D).

An alternative output function is $Z = Q_1\overline{Q}_0$.

The output function is therefore

$$Z = IQ_1Q_0$$

The logic diagram for the complete system is given in Fig. 5.12.

Fig. 5.12   Circuit diagram of the sequence detector.

---

**Exercise 5.5**   Design a sequential system which detects the bit pattern 1101 in a stream of bit-serial data.

### State minimization in sequential logic

When modelling sequential systems on a state transition diagram it is important to have as few internal states as possible. This leads to simpler and cheaper hardware.

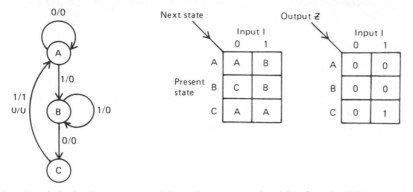

Fig. 5.13   A minimized state transition diagram and tables for the 101 sequence detector.

A state transition diagram can be minimized by combining two or more states provided their next states and outputs are identical. In the 101 sequence detector (Worked Example 5.6) comparing states A and D we see the next state is always A if I is 0, B if I is 1, and the output is always 0. States A and D therefore have identical next states and outputs and so may be combined into a single state. State D is then eliminated from the system and any flow line directed to it can be rerouted to A. The state-minimized system is shown in Fig. 5.13.

In this minimized system only three states are required. Two JKFFs are still necessary but the driving logic is simpler as the fourth state, which could be represented but is not now needed, becomes a 'don't care'.

---

Identify the states that can be minimized in the following state transition diagram:   **Worked Example 5.7**

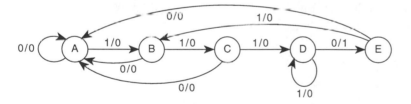

*Solution*   The present/next and output table for the above system is

| Present | Next | | Output | |
|---|---|---|---|---|
| | I = 0 | I = 1 | I = 0 | I = 1 |
| A | A | B | 0 | 0 |
| B | A | C | 0 | 0 |
| C | A | D | 0 | 0 |
| D | E | D | 1 | 0 |
| E | A | B | 0 | 0 |

By inspection, it can be seen that present states A and E have identical next states and outputs. Therefore, A ≡ E.

The system can be minimized by removing state E and replacing it with A.

**Exercise 5.6** Design a sequential system, using 2 JKFFs and driving logic, that detects the pattern 1011 in a stream of bit-serial data.

### Equivalent states

It is sometimes possible for states to be combined even if the 'next states' are not identical. The corresponding outputs must still, however, be the same for every input condition if two or more states are to be reduced. Consider the following specification for a sequential machine:

| Present state | Next state I = 0 | Next state I = 1 | Output I = 0 | Output I = 1 |
|:---:|:---:|:---:|:---:|:---:|
| A | B | C | 1 | 0 |
| B | D | E | 1 | 0 |
| C | A | F | 1 | 1 |
| D | D | E | 1 | 0 |
| E | A | F | 1 | 1 |
| F | A | A | 0 | 0 |

'Present states' D and E have been removed. Where D and E were 'next states' they have been replaced by B and C respectively.

On inspection, C and E are identical as are B and D, and the machine can be reduced to the following specification:

| A | B | C | 1 | 0 |
|:---:|:---:|:---:|:---:|:---:|
| B | B | C | 1 | 0 |
| C | A | F | 1 | 1 |
| F | A | A | 0 | 0 |

Having removed the identical states it can now be seen that A and B are identical, and B can be eliminated giving a three-state machine with the following specification:

| A | A | C | 1 | 0 |
|:---:|:---:|:---:|:---:|:---:|
| C | A | F | 1 | 1 |
| F | A | A | 0 | 0 |

In the original six-state machine, although the states A and B have been shown to be equivalent they do not lead to the same immediate next states. The final state of the machine will, however, be the same regardless whether one starts from state A or state B. One sequence of states is therefore redundant and can be removed from the machine.

States A and B are called equivalent states, and can be detected by repeated identification of identical state pairs such as B and D or C and E, followed by removal of one of the redundant states.

### The implication chart

In simple specifications identical states can be detected, and then any equivalent states located and minimized. As sequential systems become larger it is more

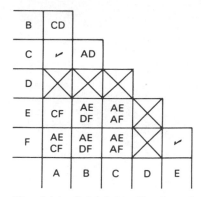

Fig. 5.14  Initial implication chart.

difficult to carry out this inspection operation. The implication chart provides a more formal approach to state minimization, and it can be incorporated into a computer program.

The implication chart shown in Fig. 5.14 takes the form of a triangular array of cells, one for every possible pair of states in the original machine. It can be drawn as a matrix of rows and columns with the rows labelled with all the states except for the first and the columns with all except the last state. Each pair of states can be described in one of three ways:

1. A pair is incompatible if the output states differ, in which case an X is entered on the chart.
2. A state pair is identical if the corresponding next state and outputs are the same for every input condition. This is denoted by a ✓ on the chart.
3. If the pair of states have identical outputs, but different next states, they may be equivalent, and the conditions which must be satisfied for this equivalence are entered on the chart.

Consider the following finite state machine:

| Present | Next state | | Output | |
|---------|-------|-------|-------|-------|
| state | $I = 0$ | $I = 1$ | $I = 0$ | $I = 1$ |
| A | E | C | 0 | 0 |
| B | E | D | 0 | 0 |
| C | E | A | 0 | 0 |
| D | D | E | 0 | 1 |
| E | A | F | 0 | 0 |
| F | A | F | 0 | 0 |

On inspection E can be seen to be identical to F and the machine can be reduced by one state. Other state reductions arising from equivalent states cannot be detected by inspection but can be located using an implication chart. State A may be equivalent to B but depends upon the equivalence between states C and D. These are the next states entered when input I is equal to 1. The condition C,D is entered into the chart for cell A,B. A ≡ C depends upon itself (A,C), and this

105

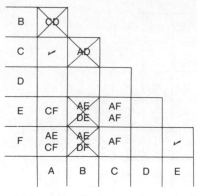

Fig. 5.15   After first pass through chart.

pair can be regarded as identical states, and a tick entered into the chart at cell A,C. A,D is an incompatible pair as the outputs are inconsistent. For state A, $Z = 0$ if $I = 1$, whereas for D, $Z = 1$ if $I = 1$. State pairs with different outputs can NEVER BE EQUIVALENT.

The first pass through the chart involves taking each incompatible state pair (this is given by the co-ordinates of the cells containing a cross) and comparing it with all the conditions contained in the chart. If any state pair depends upon the equivalence of an incompatible pair it cannot be satisfied and becomes incompatible and must be deleted from the chart. After the first pass through the chart, new incompatibles are located as shown in Fig. 5.15.

Further passes are now required to eliminate any remaining conditions dependent on the new incompatibles found during the first pass through the chart. When no more pairs of states can be deleted from the chart the co-ordinates of remaining cells are the equivalent states.

The equivalent state pairs in Fig. 5.16 are

$$A \equiv C, \ A \equiv E, \ A \equiv F, \ C \equiv E, \ C \equiv F, \ E \equiv F$$

These pairs can now be combined into larger sets providing a transitive relationship exists between all state pairs. In this example they can all be combined into a single set:

(A,C,E,F)

The incompatibles entered in the initial chart are omitted in subsequent charts – for clarity.

$A \equiv B$ and $B \equiv C$ is transitive if $A \equiv C$.

In more complex problems several sets of states may occur.

|   | A | B | C | D | E |
|---|---|---|---|---|---|
| B |  |  |  |  |  |
| C | ✓ |  |  |  |  |
| D |  |  |  |  |  |
| E | CF |  | AE AF |  |  |
| F | AE CF |  | AF |  | ✓ |

Fig. 5.16   Final chart.

106

which means that all four states are equivalent, and C, E and F can be replaced by A.

A covering of the original machine must now be obtained. Each state or its equivalent must be transferred to the reduced machine. A, C, E and F can be represented by A, but B and D, which do not combine with any other states, must be transfered to the reduced machine as single states.

The covering is therefore

(A,C,E,F),B,D

In a covering all the original states or their equivalents must be present.

and the reduced set of states is A,B,D.

By eliminating equivalent states C, E and F, we can reduce the specification to a three-state machine:

| Present | Next state | | Output | |
|---------|:-------:|:-------:|:-------:|:-------:|
| state | I = 0 | I = 1 | I = 0 | I = 1 |
| A | A | A | 0 | 0 |
| B | A | D | 0 | 0 |
| D | D | A | 0 | 1 |

which can then be implemented with two flip-flop devices using standard sequential design techniques.

### Asynchronous sequential design

The design strategies for synchronous sequential logic where the switching is controlled by a clock and asynchronous systems where state transitions are brought about by changes in the input data are in principle very similar. In practice, however, owing to finite propagation delays of indeterminate length in practical circuits, unforeseen behaviour can arise which may lead to circuit malfunctions in asynchronous systems.

Suppose we wish to design a sequential system which has two inputs A and B, and two outputs S and W. The system must operate solely on the input sequence and S is to be set to 1 after the sequence 00, 01, 00, 10 has been input on AB. If any different sequence occurs on AB the second output W must be set to remain 1. The device can be regarded as a simple sequential lock with the lock opening if S = 1 and an alarm sounding if W = 1.

A state transition diagram can be built up from this device. As there is a likelihood of timing problems it is better to drive the output from the state variables only. This known as a Moore model, and each node contains state and output data. The directed lines are only labelled with the inputs. The complete state transition diagram is given in Fig. 5.17.

Compare with the Mealy model where the outputs depend on both state and input variables.

When the correct sequence on AB is applied the system switches from state ① through ② and ③ to ④, and the output S is then set to 1. Any deviation in the correct sequence would lead to state ⑤ with output W at 1 and all further inputs would keep the system in this state. We will simplify the problem by assuming that no further inputs are applied after state ④ when the 'lock opens'.

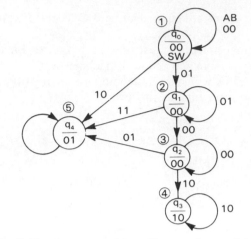

Note: if the inputs are not synchronized, certain consecutive inputs where both bits change value cannot occur, e.g. AB = 11 cannot be applied when in state ①.

Fig. 5.17  State transition diagram for an 'electronic lock'.

A sling is a directed line on the STD which leaves and then re-enters the same state.

Two important differences can be noted compared with synchronous systems. Each node must have a 'sling' with an input to hold it in that particular state. State transitions can only be enabled by input changes. Asynchronous systems cannot therefore differentiate between consecutive input data which is the same. If there is no timing data present, input sequence AB = 00,01,01,00 cannot be distinguished from AB = 00,01,00.

Imagine A and B being driven from hand-operated switches.

The second difference is that each node cannot respond to all input values. For example, input AB = 00 holds the system in state ①; 01 or 10 causes transitions. The input AB = 11 cannot be applied directly to state ① as both variables would have to change from 0 to 1 simultaneously. In an asynchronous system this cannot happen instantaneously. Any delay between the two inputs will cause intermediate values AB = 01 or AB = 10 to occur. These intermediate values will cause either state ② or state ⑤ to be entered. In general if an input is latching a machine into a given state, the only allowable inputs that can be applied must differ by only 1 bit from the latching input. The inputs which cannot occur can be regarded as 'don't care' conditions. The full flow table can be compiled from the state transition diagram and for the lock example is given below:

<table>
<tr><td rowspan="2"></td><td rowspan="2"></td><td colspan="4">Inputs AB</td></tr>
<tr><td>00</td><td>01</td><td>11</td><td>10</td></tr>
<tr><td rowspan="5">Internal state</td><td>$q_0$</td><td>①</td><td>2</td><td>d</td><td>5</td></tr>
<tr><td>$q_1$</td><td>3</td><td>②</td><td>5</td><td>d</td></tr>
<tr><td>$q_2$</td><td>③</td><td>5</td><td>d</td><td>4</td></tr>
<tr><td>$q_3$</td><td>d</td><td>d</td><td>d</td><td>④</td></tr>
<tr><td>$q_4$</td><td>⑤</td><td>⑤</td><td>⑤</td><td>②</td></tr>
</table>

2 is unstable.
② is stable.
d is 'don't care'.

In the initial flow table each circuit state has its own internal state, and hence each row contains only one stable state, although that state can occur more than once as is the case with ⑤. The user can only change inputs A and B, which give rise to horizontal motion in the flow table. Changes between internal states can be achieved by unstable to stable transitions in the logic. The output depends only on the internal state and output S is set to 1 when in $q_3$ (stable state ④), and W is set to 1 when in $q_4$ (any of states ⑤).

108

## State merging

Some state minimization can be achieved by merging rows on the flow table. Two or more rows can be merged providing there is no conflict between the corresponding states. An unstable state can be absorbed by its own stable state, and 'don't care' conditions can be reassigned. Dissimilar states cannot be combined regardless of their stability conditions.

The two internal states that can be merged in the lock problem are $q_2$ and $q_3$, and the following flow table results:

$q_3$ – the merged state – is removed.

|  | Inputs AB | | | |
|---|---|---|---|---|
|  | 00 | 01 | 11 | 10 |
| $q_0$ | ① | 2 | d | 5 |
| $q_1$ | 3 | ② | 5 | d |
| $q_2$ | ③ | 5 | d | ④ |
| $q_4$ | ⑤ | ⑤ | ⑤ | ⑤ |

Internal state

Whilst merging reduces the number of internal states it may be at the expense of more complicated output logic. States ③ and ④ have different outputs which when merged cannot be derived from the internal state variables alone. The output functions must now depend on both the internal state and the input variables, and will inevitably become more complex, although the number of internal states and their driving logic will be reduced.

## State assignment

In synchronous sequential machines any state assignment can be made without encountering malfunctions in the logic. This is because the present to next state transition is determined by a controlling clock, and providing the longest delay path through the logic is less than one clock period all the desired conditions will be set up on the flip-flop inputs by the time the clock edge arrives. Correct switching will then occur. If there is no controlling clock and the system has more than one state variable, there is a possibility of unequal delays as information passes through different paths in the logic. This can be particularly serious if a particular state transition requires two or more state variables to change at the same instant. Suppose at time $t$ the system is in internal state $Q_1 Q_0 = 00$ and the next state is required to be 1. $\overline{Q}_1$ and $Q_0$ are generated by different logic functions, and, as it is virtually impossible for them to both change from 0 to 1 within a few nanoseconds, the intermediate states 01 or 10 will be entered. This can result in the system entering incorrect circuit states. If two or more state variables are required to switch simultaneously, a RACE condition is said to exist. Races can be critical or non-critical. In a critical race there is a possibility of the machine entering incorrect states. Critical races must therefore be avoided at all costs. If the race is non-critical the logic will always enter the correct state, although it may take different state paths depending on the delays. Non-critical races can be tolerated.

The problem concerning state assignments and the occurrence of races can be illustrated in the lock problem. Let the state assignments $q_0 = 00$, $q_1 = 01$,

Some state assignments may lead to marginally simpler logic.

This problem has already been encountered with the input variables.

109

$q_2 = 10$ and $q_4 = 11$. The merged flow table with these assignments becomes

|  | | Inputs AB | | | |
|---|---|---|---|---|---|
|  | $Q_1Q_0$ | 00 | 01 | 11 | 10 |
| State | 0 0 | ① | ② | d | 5 |
|  | 0 1 | 3 | ② | 5 | d |
|  | 1 0 | ③ | 5 | d | ④ |
|  | 1 1 | ⑤ | ⑤ | ⑤ | ⑤ |

$Q_i$ is a state variable.
$q_i$ is a state label.

With this state assignment a critical race occurs when the input is $AB = 00$ and the system is required to switch from unstable 3 to stable ③. The internal state transition is 01 to 10 which requires both $Q_1$ and $Q_0$ to change instantaneously. If $Q_1$ has a greater delay than $Q_0$, internal state 11 will occur and the system will enter stable state ⑤. If the delay times are reverse and we momentarily enter internal state 00 the circuit will latch into state ①. Another critical race occurs from 5 to ⑤ when the input is $AB = 10$.

An alternative state assignment must be found which avoids these critical races. We could try $q_0 = 00$, $q_1 = 01$, $q_2 = 11$ and $q_4 = 10$.

|  | | Inputs AB | | | |
|---|---|---|---|---|---|
|  | $Q_1Q_0$ | 00 | 01 | 11 | 10 |
|  | 0 0 | ① | 2 | 5 | 5 |
|  | 0 1 | 3 | ② | 5 | 5 |
| State | 1 1 | ③ | 5 | 5 | ④ |
|  | 1 0 | ⑤ | ⑤ | ⑤ | ⑤ |

The previous critical races do not occur with this state assignment, but a new race now exists when $AB = 11$. The system will enter one of the 'don't care' states when trying to switch from unstable 5 to stable ⑤. This problem is overcome by making the 'don't care' state into unstable 5. Hence the system will then finally enter stable ⑤ regardless of the state path. This race has now become non-critical and it can be tolerated within the design. In general it is always best to avoid 'don't care' conditions in asynchronous design as they can lead to anomalous behaviour if any unforeseen delays occur and these states are inadvertently entered.

The final assigned critical race-free flow table is shown below.

|  | | Inputs AB | | | |
|---|---|---|---|---|---|
|  | $Q_1Q_0$ | 00 | 01 | 11 | 10 |
|  | 0 0 | ① | ② | d | 5 |
|  | 0 1 | 3 | ② | 5 | d |
| State | 1 1 | ③ | 5 | d | ④ |
|  | 1 0 | ⑤ | ⑤ | ⑤ | ⑤ |

## The asynchronous next state and output logic

Two approaches to asynchronous design can be followed. Either we can use combinational logic with feedback or the design can be based on set–reset flip-flops.

(a) *Feedback logic*

The block diagram for the next state logic using feedback is given in Fig. 5.18.

$Q_i'$ is $Q_i$ after being fed back to an input.

Two logic functions $Q_1$ and $Q_0$ are required. Both are functions of the four variables A, B, $Q_1'$ and $Q_0'$.

The logic can be derived from the flow table and built up on K-maps by applying two important rules:

1. If the circuit is in a stable state the output $Q_1$ and $Q_0$ must be set to the values of $Q_1'$ for that stable state.
2. If the circuit is in an unstable state $Q_1$ and $Q_0$ must take the values of $Q_1'$ and $Q_0'$ for the NEXT stable state the circuit is to enter.

The K-maps for $Q_1$ and $Q_0$ are

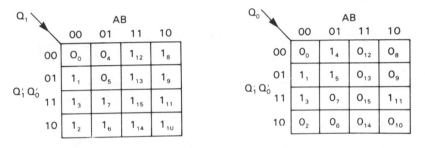

Cell 0 on both K-maps corresponds to state ①. It is stable with $Q_1'Q_0'$ assigned the value 00. Hence both $Q_1$ and $Q_0$ must be 0 and these values are entered in the K-map. Cell 1 corresponds to unstable state 3. This must lead to stable state ③ where $Q_1'Q_0' = 11$, so both $Q_1$ and $Q_0$ are therefore set to logic level 1 in cell 1. By considering every state on the flow table the K-maps can be completed and the logic functions $Q_1 = f(ABQ_1'Q_0')$ and $Q_0 = f(ABQ_1'Q_0')$ evaluated. In their minimal form they are

Refer to the final flow table on page 110.

$$Q_1 = A + Q_1' + \overline{B}Q_0'$$
$$Q_0 = \overline{AB}\overline{Q}_1' + \overline{AB}Q_0' + \overline{B}Q_1'Q_0'$$

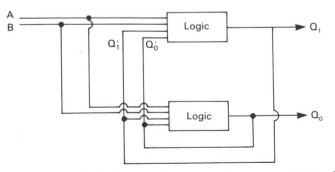

Fig. 5.18   Block diagram from asynchronous next state logic.

The output logic requires $S = 1$ when in state ④ and $W = 1$ when in state ⑤. By inspection of the flow table the required output logic functions are

$$S = A\overline{B}Q_1Q_0 \quad \text{and} \quad W = Q_1\overline{Q_0}$$

In order to illustrate the principles of asynchronous design this problem has been reduced to its simplest form. If the circuit were to be built or simulated, the designer would have to provide some mechanism to reinitialize the machine because there is at present no means to return the machine to state ① from either of the final states ④ or ⑤. The simplest reset strategy is to AND both next state logic outputs $Q_1$ and $Q_0$ with a reset control line R. A logical 0 on R will then reset the machine and $R = 1$ will enable the machine.

(b) *Set–reset flip-flop based design*
If SRFFs are to be used the driving logic has to be designed. The block diagram for the system is shown in Fig. 5.19.

Fig. 5.19    SRFF-based 'next state' logic.

Four driving functions are required to excite the flip-flops. The state transition table for an SRFF is

Q is 'present' state.
Q + d$t$ is 'next' state.

Flow table is
AB

|  | 00 | 01 | 11 | 10 |
|---|---|---|---|---|
| $Q_1Q_0$ 00 | ① | 2 | 5 | 5 |
| 01 | 3 | ② | 5 | 5 |
| 11 | ③ | 5 | 5 | ④ |
| 10 | ⑤ | ⑤ | ⑤ | ⑤ |

For example:

$2 \equiv q_1$  and
② $\equiv q_2$

The 'present' state is $q_1$ and 'next' state $q_2$.

Refer to SRFF transition table.

| S | R | $Q_t \to Q_t + \mathrm{d}t$ |
|---|---|---|
| 0 | d | $0 \to 0$ |
| 1 | 0 | $0 \to 1$ |
| d | 0 | $1 \to 1$ |
| 0 | 1 | $1 \to 0$ |

The driving logic is obtained with reference to the flow table. The following rules apply:

1. If the circuit is in a stable state the 'next' internal state is the same as the 'present' internal state.
2. If the circuit is in an unstable state the 'present' internal state is the unstable state and the 'next' internal state is the subsequent stable state the circuit enters.

As an example of the use of those two rules consider state 1 on the flow table. The present state is $Q_1'Q_0' = 00$ and the next state must be $Q_1'Q_0' = 00$. The transitions on the state variables are therefore

$$Q_1' = 0 \to 0 \quad \text{and} \quad Q_0' = 0 \to 0$$

requiring $S_1 = 0$, $S_0 = d$, $R_1 = 0$ and $R_0 = d$.

Now consider unstable state 3. The present internal state is 01 and the next internal state must be 11 corresponding to stable state ③. State variable transitions are $Q_1 = 0 \rightarrow 1$ and $Q_0 = 1 \rightarrow 1$ requiring $S_1 = 1$, $R_1 = 0$, $S_0 = d$ and $R_0 = 0$.

The complete K-maps for the flip-flop driving functions are

$S_1$, rows $Q_1Q_0$, columns AB:

| $S_1$ \ AB | 00 | 01 | 11 | 10 |
|---|---|---|---|---|
| 00 | . | . | 1 | 1 |
| 01 | 1 | . | 1 | 1 |
| 11 | d | d | d | d |
| 10 | d | d | d | d |

$R_1$, rows $Q_1Q_0$, columns AB:

| $R_1$ \ AB | 00 | 01 | 11 | 10 |
|---|---|---|---|---|
| 00 | d | d | . | . |
| 01 | d | d | . | . |
| 11 | . | . | . | . |
| 10 | . | . | . | . |

$S_0$, rows $Q_1Q_0$, columns AB:

| $S_0$ \ AB | 00 | 01 | 11 | 10 |
|---|---|---|---|---|
| 00 | . | 1 | . | . |
| 01 | d | d | . | . |
| 11 | d | . | . | d |
| 10 | . | . | . | . |

$R_0$, rows $Q_1Q_0$, columns AB:

| $R_0$ \ AB | 00 | 01 | 11 | 10 |
|---|---|---|---|---|
| 00 | d | . | d | d |
| 01 | . | . | 1 | 1 |
| 11 | . | 1 | 1 | 1 |
| 10 | d | d | d | d |

giving the following driving logic functions:

$$S_1 = A + \overline{B}Q_0 \qquad R = 0$$
$$S_0 = \overline{AB}\overline{Q_1} \qquad R_0 = A\overline{Q_1} + BQ_1$$

The output logic functions would of course be the same as those for the design based on combinational logic with feedback.

Asynchronous logic is potentially very fast as it operates as soon as the data is available and does not have to wait for clock signals. It is, however, prone to timing problems which constrain the design options. Strategies for removing critical races are essentially trial and error and therefore difficult to automate. In some machines it may be impossible to assign states which are critical race free. Under these conditions additional internal states must be introduced to the flow table. Initially they will have 'don't care' circuit states, but with suitable state assignments the machine can be routed through these dummy states and the 'don't care' reallocated in order to switch the system to the desired stable state without malfunction.

It is advisable to exercise asynchronous logic designs on a logic simulator as unforeseen timing problems can arise. In general, unless there are compelling reasons for using asynchronous logic, the synchronous option should always be chosen because it is inherently more reliable and also easier to design.

**Exercise 5.7** Design an asynchronous logic system which will output logical 1 whenever consecutive 2-bit inputs are in ascending Gray code (the 2-bit Gray code cycle is 00, 01, 11, 10, 00, etc.).

### Algorithmic state machines (ASMs)

When designing large sequential systems with many input and internal states, the state transition diagram approach becomes very cumbersome and complex. An ASM chart provides a simpler block diagram representation of the behaviour of the finite state machine and has many similarities in its structure to the flow charts used in software specification and design. The chart has three components as shown in Fig. 5.20.

a) State box        b) Decision box        c) Conditional output box

Fig. 5.20   Components of an ASM chart.

The use of ASM charts is best illustrated by example. Consider the '101' sequence detector of Worked Example 5.6. An ASM chart for this machine is given in Fig. 5.21. The top state box represents the start. On leaving the start state a decision box is entered which tests the input I. If $I = 0$ then the start state is re-entered. If $I = 1$ we proceed to the second state box which represents the condition 'first bit of sequence received'. The output specified within the box remains at 0 as the complete input sequence has not yet been received. The reader can verify for him- or herself that the system only enters the 'sequence correct' state when 101 has been received on the input. Any deviation causes the system either to restart or to enter the appropriate state according to the value of the input.

In general, ASMs are easier to read than state transition diagrams. Decision boxes can be used to test logic functions and therefore each state box requires only one exit path compared with the STD method where $2^n$ directed lines are required at each node when there are $n$ input variables to the system.

It is, however, much more difficult to minimize the logic on an ASM chart, as all the input transitions are not explicitly shown but implied by the decision boxes. The chart of Fig. 5.21 has a single output state associated with every internal state. This can give rise to a larger number of internal states than if the output depended on both internal states and input variables. If conditional output boxes are introduced after decision boxes, an output dependence on both state and input variables can be defined. In a synchronous machine, switching between state boxes is controlled by the systems clock.

**Reading an ASM chart**

The specification tables for the next state can be obtained by noting for each state the input required to take it to its next state. In Fig. 5.20 state 00 and input $I = 1$ lead to state 01, whereas $I = 0$ takes the machine back to state 00. The full table is given on page 116.

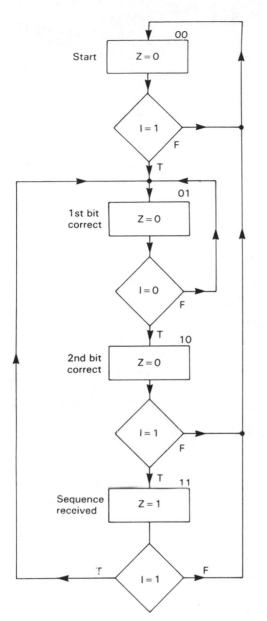

Fig. 5.21   An ASM chart for a '101' sequence detector.

| Present state $Q_1'Q_0'$ | | Input $I$ | Next state $Q_1Q_0$ | | Output $Z$ |
|---|---|---|---|---|---|
| 0 | 0 | 0 | 0 | 0 | 0 |
| 0 | 0 | 1 | 0 | 1 | 0 |
| 0 | 1 | 0 | 1 | 0 | 0 |
| 0 | 1 | 1 | 0 | 1 | 0 |
| 1 | 0 | 0 | 0 | 0 | 0 |
| 1 | 0 | 1 | 1 | 1 | 1 |
| 1 | 1 | 0 | 0 | 0 | 0 |
| 1 | 1 | 1 | 0 | 1 | 0 |

When designing the machine it is simpler to use D-type flip-flops as the inputs to the DFFs are simply the values of the outputs in the 'next' state and can be read directly from the specifications table. In the above example the driving functions $D_1$ and $D_0$ for flip-flops generating state bits $Q_1Q_0$ are

$$D_1 = \overline{Q_1'}Q_0'\overline{I} + Q_1'\overline{Q_0}I$$
$$D_0 = \overline{Q_1'}\,\overline{Q_0'} + \overline{Q_1'}Q_0I + Q_1'\overline{Q_0'}I + Q_1'Q_0'I$$

and the output function is

$$Z = Q_1Q_0$$

In general, ASMs enable large sequential systems to be specified and designed. They do not, however, lead directly to minimal design, and it can also be quite difficult to check the validity and completeness of an ASM chart in the same way that it is difficult to prove a software system from its flow chart. It is important to realize that whilst software systems can be easily modified changes are not so simple to carry out in hardware. The use of software simulation of hardware therefore has an increasingly important role to play in the design of logic systems. This provides a facility whereby designs can be verified before being committed to silicon, whether in the form of discrete gates or customized integrated circuits.

For further reading on ASM techniques see Lewin, D. and Protheroe, *Design of Logic Systems*, (2nd edn Chapman & Hall, 1992).

### Summary

A general sequential circuit contains two distinct logic functions: the next state logic and the output logic. Three sets of data can be identified: $I$ represents the inputs, $Q$ the internal states, and $Z$ the outputs.

In many simple sequential systems the outputs are identical to the internal states and no output logic is required. Sequential logic can be either synchronous or asynchronous. In asynchronous systems, timing and propagation delays can cause malfunctioning, although the circuits will operate at high speed. When designing sequential logic using edge-triggered flips-flops, suitable falling edges must be identified within the system to give the desired switching sequences. In some circumstances these edges may not exist or cannot be identified and output logic is then required to encode the state sequence of a counter which can be built.

The basic building block for discrete synchronous logic is the JKFF. The present/next state table provides a suitable means of specifying the state sequences within a system. The driving logic for the JKFF can then be obtained from the JK transition table and the system's present/next state table. The inputs to the driving functions and the output logic are the state variables and the external inputs.

An autonomous sequential system has no external inputs and if the controlling clock is replaced by a data source, the JKFF can operate on asynchronous data. There is no separate clock in these systems and the logic operations are synchronized by the inputs.

The state transition diagram is a powerful tool, enabling a system to be represented graphically. This allows the properties of a sequential system to be identified and scrutinized. The present/next state table can then be deduced from the transition diagram.

Sequential systems can be state minimized. If two or more states lead to identical next states and have consistent outputs they may be combined into a single state. The complexity of the next state logic can therefore be reduced.

For large complex systems, computer-aided design may be necessary. In practice, owing to the cost of hardware being a small item in any design programme, the minimization of sequential logic may not be cost effective. In VLSI design, however, it could be advantageous to minimize the logic. If the basic cell of an iterative system is designed in its simplest form the total number of cells that can be fabricated on a given area of silicon is increased. Logic minimization may therefore lead to more powerful single-chip systems.

An implication chart can be used to state-minimize sequential logic and locate equivalent states which cannot readily be identified on the specification tables. The method can be computerized.

Asynchronous logic can be designed using flow tables. These systems are, however, prone to timing problems which restrict the input sequences and give rise to race conditions in the state transitions. Critical races will lead to malfunctions and must always be avoided. Non-critical races do not affect the overall behaviour of the system and can therefore be tolerated.

In general asynchronous logic can operate at high speed but if possible should be avoided because of inherent timing problems.

Algorithmic state machine (ASM) charts which are similar in structure to software flow charts can be used to specify and model large sequential systems.


## Problems

5.1   Design a binary counter using asynchronous TFFs and driving logic which will count in 8421 BCD a series of randomly occurring pulses on its input I.

5.2   Design a 4-bit walking code generator, using a 3-bit counter and output logic. (Walking code is 0000, 0001, 0011, 0111, 1111, 1110, 1100, 1000, 0000, etc.) Redesign the system using shift registers and compare the circuits.

5.3   Using JKFFs, design a system that will output the following 2-bit code: 00, 01, 11, 10 and then continuously repeat the sequence from 01. Repeat the design using (i) TFFs and (ii) DFFs.

5.4   Design a logic system to operate a set of traffic lights. The lights must be switched on in the following sequence: red, red and amber, green, amber, red, etc., and a logical 1 from the controller switches a light on. The system is controlled by the following waveform:

Red

Red and amber

Green

Amber

Red

5.5 Design the logic for a 3-bit up/down counter which counts up when the input I is 1 and down when I is 0. The counter should change by one state per clock pulse and DFFs should be used in the design.

5.6 A faulty batch of JKFFs have the following state transition diagram:

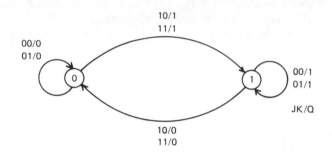

Obtain the transition table for the faulty flip-flops. How should the input terminals be relabelled so that the devices could be sold as standard flip-flops? What type would they be?

5.7 Design a logic system to implement the following state transition diagram:

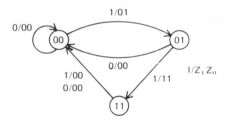

Describe in words, the behaviour of the circuit.

5.8 Design a system to indicate on output Z, whether, in a string of 2-bit inputs, the current input is greater than the previous input (Z = 1) or not (Z = 0).

5.9 State-minimize the following sequential machine:

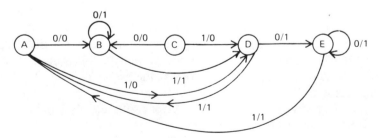

5.10 Design a synchronous sequential logic system, using the minimum number of JKFFs together with NAND gates for driving and output logic, that will provide an output $Z = 1$, as long as the repetitive sequence AB = 00, 01, 11, 10, 00, etc., is applied to the inputs. (One 2-bit word is input on each clock pulse.) If any input is applied out of sequence, the output must go to 0. The output should become 1 again only if the next correct input is applied.

5.11 Design an entry system that provides a signal to unlock a door provided the input sequence AB = 00, 10, 01, 11 is applied and sounds an alarm continuously if any deviation occurs on the inputs. Comment on any possible timing problems.

5.12 Design a machine that compares pairs of 4-bit numbers input bit-serially on two lines A and B, and outputs a signal whenever two words are identical.

5.13 Use an implication chart to reduce the following machine:

| | Next state | | |
|---|---|---|---|
| Present state | $I = 0$ | $I = 1$ | Output |
| 1 | 1 | 5 | 1 |
| 2 | 5 | 3 | 0 |
| 3 | 6 | 2 | 1 |
| 4 | 7 | 6 | 0 |
| 5 | 1 | 3 | 0 |
| 6 | 8 | 7 | 1 |
| 7 | 4 | 8 | 0 |
| 8 | 6 | 2 | 1 |

5.14 Obtain a state diagram and flow table for an asynchronous logic system whose output Z is 1 provided consecutive inputs AB are in ascending Gray code, otherwise $Z = 0$.

# 6 The digital system

**Objectives**
- [ ] To examine the influence of integrated circuit technology on logic design.
- [ ] To investigate the concept of programmable logic.
- [ ] To design logic systems based on memories and arrays.
- [ ] To specify a system as a set of registers and its operation as a series of register transfers.
- [ ] To examine the bus concept.
- [ ] To introduce the basic principles of fault diagnosis in a digital circuit

The development of integrated circuit technology has had a profound influence on logic design. In the 1950s all logic gates were built with discrete components – transistors, resistors, capacitors and the like. The advent of the transistor, a decade before, had made the construction of individual gates a practical proposition. The late 1960s saw the introduction of very small scale integrated circuits containing less than 100 electronic components per chip. A typical product that appeared in the 1960s is the quad two-input NAND package – a 14 pin dual in-line integrated circuit that contains four NAND gates, each having two inputs. These circuits enable logic systems to be constructed from discrete gates.

Since the introduction of integrated circuits in the 1960s, dramatic developments have taken place. The complexity has, on average, doubled every three years and we now have very large scale devices that contain the equivalent of 100 000 components. The cost per elementary logic function, however, has halved, in real terms, every year, giving a cost reduction of 1 million times in 20 years. This can best be illustrated by considering memory. In 1965, 1 bit of semiconductor memory comprised a DFF made out of discrete components. The cost would have been the equivalent of a graduate engineer's pay for 20 hours' work. Today, 1 bit of memory is one-quarter-millionth of a 256k-bit memory chip and the cost of the whole chip is less than $100. The bit cost is equivalent to an engineer's pay for a fraction of a second.

A wide range of dedicated logic integrated circuits are now available. The reader is directed to the manufacturers' catalogues for further details. In this chapter we will be examining integrated circuits that are not restricted to single-function operations and that can be used in the design of a general digital system.

## Programmable logic

A dedicated logic circuit will only perform the function it was designed to implement. A combinational circuit behaves according to its truth table, whereas a sequential circuit's characteristics are governed by its state table. In a programmable logic system, a standard hardware structure is electronically reconfigurable so that the different logic operations can be performed, without having to change the circuit physically.

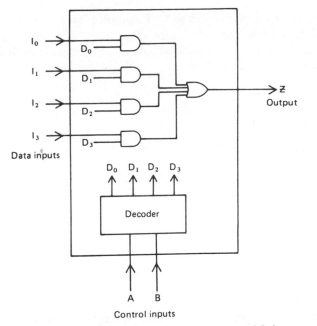

Fig. 6.1   Logic diagram of a 4 to 1 multiplexer.

*The multiplexer*

The multiplexer is a multi-input, single-output circuit whose prime function is to convert parallel data to serial data. It differs from the parallel input/serial output shift register in one important respect – it has additional control inputs that address individual data input lines. The inputs can therefore be output in any order dependent on the control address.

Figure 6.1 shows a 4 to 1 multiplexer. The control inputs must be able to address any input line; hence two control inputs are required by this multiplexer.

The control inputs enter a decoder that has four outputs. Each responds to a specific input value on AB giving

$$D_0 = \overline{A}\,\overline{B}$$
$$D_1 = \overline{A}\,B$$
$$D_2 = A\overline{B}$$
$$D_3 = A B$$

Hence for a given control input, only one decoder output will be at logical 1. The decoder outputs are ANDed with corresponding data inputs $I_i$ and then ORed together to give the multiplexer output Z. Thus the output Z takes on the value of $I_i$, where $i$ is the address AB, applied to the control inputs. Hence the multiplexer is a parallel-to-serial data converter.

The multiplexer can also be used as a universal combinational logic function. Any logic function of $n$ variables can be set up in a $2^n$ to 1 multiplexer. The control inputs become the variables of the logic function and the data inputs are driven from logical switches. The truth table of the function to be programmed into the multiplexer must be evaluated and each minterm which requires an output set to 1 identified. The logical switches which are addressed by the minterms are set to

If the addresses are driven from a pure binary counter, the multiplexer acts as a shift register. It does not, however, have any internal storage; therefore the inputs must be held constant during operation.

1 and all other data inputs remain at 0. The multiplexer will then perform the combinational function as specified by the truth table. The same multiplexer can be programmed to perform another combinational function simply by changing the setting of the data input switches.

**Worked Example 6.1**  Implement an EXCLUSIVE OR function in a 4 to 1 multiplexer. Reprogram the multiplexer to perform the AND function.

*Solution*  The truth table for EX.OR is

| AB | F |
|----|---|
| 00 | 0 |
| 01 | 1 |
| 10 | 1 |
| 11 | 0 |

Minterms of EX.OR are $A\overline{B}$ and $\overline{A}B$. If A and B are applied to the control inputs of the multiplexer, data inputs $I_1$ and $I_2$ will be selected by the minterms. These inputs must be set to 1.

Hence the multiplexer circuit needed to implement EX.OR is

For the AND function, we only have one minterm (AB). Therefore data input $I_3$ must be set to 1 giving

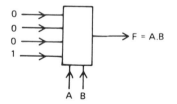

The 16 functions are the binary connectives (see Chapter 2).

$n$ inputs, $2^n$ input states, $2^{2^n}$ different output sets.

A 10-variabled function would require a single multiplexer with 1024 ($2^{10}$) data inputs. This is physically impossible.

A $2^n$ to 1 multiplexer can be programmed to perform any function of $n$ variables. This means that a 4 to 1 multiplexer can implement 16 functions of two variables, and with a 16 to 1 multiplexer we can perform any one of the 65 536 possible functions of four variables. There is, however, a physical limitation of the number of data inputs that are available on an integrated circuit, and this will limit the complexity of the combinational functions that may be programmed into the multiplexer. It is possible to overcome this restriction by allowing either logical variables or logic functions to be input to the data lines. The latter can be developed as a multilevel multiplexer system.

Consider the three-variabled functions

$$F(ABC) = \Sigma(1,2,4,6) \tag{6.1}$$

which we wish to implement in a 4 to 1 multiplexer. This can be achieved by making one variable available (say A) to the data inputs, and B and C to the control inputs.

The full Boolean equation of the function is

$$F = \overline{A}\,\overline{B}C + \overline{A}B\overline{C} + A\overline{B}\,\overline{C} + AB\overline{C} \tag{6.2}$$

Equation 6.2 must be rearranged so that it contains all possible terms of BC, the variables being applied to the control inputs. Thus

$$F = (A)\overline{B}\,\overline{C} + (\overline{A})\overline{B}C + (A + \overline{A})B\overline{C} + (0)BC$$

$$F = (A)\overline{B}\,\overline{C} + (\overline{A})\overline{B}C + (1)B\overline{C} + (0)BC \tag{6.3}$$

The bracketed terms are called residues and can be logical constants or the variable A or the inverse $\overline{A}$. Variables B and C are applied to the control inputs of the multiplexer and the residues to the data inputs. The multiplexer circuit to implement Equation 6.1 is therefore

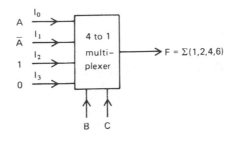

and a 4 to 1 multiplexer implements a three-variabled function.

Program an 8 to 1 multiplexer to perform the function

$$F = \Sigma(1,2,4,7,8,9,14,15)$$

## Multilayer multiplexer systems

In a multilayer multiplexer system, logic functions are generated by multiplexers, which are then input to a second and subsequent layers of multiplexers. The size of the multiplexers and the number of layers will depend on the complexity of the function to be performed. In the following analysis we will confine ourselves to a two-layer system of 4 to 1 multiplexers which can be programmed for five-variabled functions (Fig. 6.2). The methodology may be extended for larger systems.

The function F(VWXYZ) is first expanded universally about two variables (say Y and Z). The residues provide the data inputs to $M_1$. Each residue is then expanded about two variables (say W and X) and its residues, which form the inputs to the first layer, determined. The multiplexers in the first layer all have common control inputs. Hence the function can be set up by programming the first-layer data inputs with the appropriate settings (0, 1, V or $\overline{V}$) and applying the other four function variables to the control inputs of the multiplexers.

Terms are $\overline{B}\,\overline{C}$, $\overline{B}C$, $B\overline{C}$ and BC.

$(A + \overline{A}) = 1$.

If the function does not have minterms containing all combinations of B and C (e.g. BC), then the corresponding residue is 0 ((0)BC), giving a 0 output.

If the data inputs are restricted to 0 and 1, the 4 to 1 multiplexer can only act as a two-variabled function.

**Exercise 6.1**

See Fig. 6.2

123

Fig. 6.2  A 2-layer system of 4 to 1 multiplexers.

---

**Worked Example 6.2**  Implement the five-variabled function

$$F = \Sigma(1,2,4,5,7,8,12,17,19,21,28)$$

in a two-layer system of 4 to 1 multiplexers.

*Solution*  The equation of the function is

$$
\begin{aligned}
F &= \overline{V}\,\overline{W}\,\overline{X}\,\overline{Y}\,Z + \overline{V}\,\overline{W}\,\overline{X}\,Y\,\overline{Z} + \overline{V}\,\overline{W}\,X\,\overline{Y}\,\overline{Z} + \overline{V}\,\overline{W}\,X\,\overline{Y}\,Z + \overline{V}\,\overline{W}\,X\,Y\,Z \\
&\quad + \overline{V}\,W\,\overline{X}\,\overline{Y}\,\overline{Z} + \overline{V}\,W\,X\,\overline{Y}\,\overline{Z} + V\,\overline{W}\,\overline{X}\,\overline{Y}\,Z + V\,\overline{W}\,X\,Y\,Z \\
&\quad + V\,\overline{W}\,X\,\overline{Y}\,Z + V\,W\,X\,\overline{Y}\,\overline{Z}
\end{aligned}
$$

Expanding F about YZ gives

$$F = (F_1)\overline{Y}\,\overline{Z} + (F_2)\overline{Y}\,Z + (F_3)Y\,\overline{Z} + (F_4)Y\,Z$$

where

$$
\begin{aligned}
F_1 &= (\overline{V}\,\overline{W}\,X + \overline{V}\,W\,\overline{X} + \overline{V}\,W\,X + V\,W\,X) \\
F_2 &= (\overline{V}\,\overline{W}\,\overline{X} + \overline{V}\,\overline{W}\,X + V\,\overline{W}\,\overline{X} + V\,\overline{W}\,X) \\
F_3 &= (\overline{V}\,\overline{W}\,\overline{X}) \\
F_4 &= (\overline{V}\,\overline{W}\,X + V\,\overline{W}\,X)
\end{aligned}
$$

Expanding $F_1$, $F_2$, $F_3$ and $F_4$ about WX gives

$$
\begin{aligned}
F_1 &= (0)\overline{W}\,\overline{X} + (\overline{V})\,\overline{W}\,X + (\overline{V})\,W\,\overline{X} + (1)\,W\,X \\
F_2 &= (1)\,\overline{W}\,\overline{X} + (1)\,\overline{W}\,X + (0)\,W\,\overline{X} + (0)\,W\,X \\
F_3 &= (\overline{V})\,\overline{W}\,\overline{X} + (0)\,\overline{W}\,X + (0)\,W\,\overline{X} + (0)\,W\,X \\
F_4 &= (V)\,\overline{W}\,\overline{X} + (\overline{V})\,\overline{W}\,X + (0)\,W\,\overline{X} + (0)\,W\,X
\end{aligned}
$$

The multilayer multiplexer net will therefore be

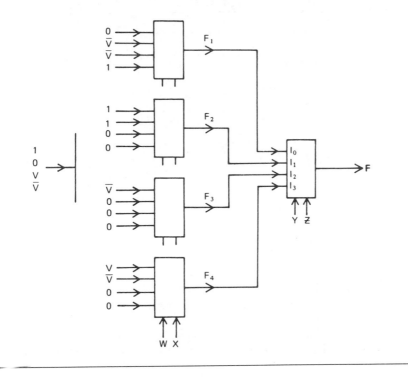

## The memory element

The memory element is a dual-purpose device. Its primary role is that of an electronic store for binary data, and as such was discussed in Chapter 2. Memories, however, are playing an increasingly important role as programmable logic devices.

A memory circuit contains several storage elements that can be addressed by a decoder circuit. The functional diagram of a 4-bit random access memory is shown in Fig. 6.3. The decoder is functionally identical to the multiplexer decoder and the storage elements are DFFs. To write into the memory the data is applied to the 'data-in' terminal which is connected to all the inputs of every DFF. The address at which the data is to be stored is supplied to inputs A and B. This will set one, and only one, of the decoder outputs to 1. Suppose we wish to write 1 into location 0. Address AB = 00 must be applied to the address inputs and decoder output $D_0$ will be set to 1. The read/write control must also be set to 1 and the inputs to AND gate 1 will both be high, giving a 1 on the output which drives the clock of DFF0, causing the logical value on 'data-in' to be entered into the flip-flop DFF0. All the remaining flip-flop decoders will be at 0 and cannot therefore store 'data-in'.

To read from the memory, the read/write control is set to 0 and inverter 9 will transmit a logical 1 to all the output AND gates (5 to 8) each of which also has as inputs a DFF output and its decoded address. The AND gate outputs are ORed together to produce the memory output whose value is the contents of the flip-flop selected by the address decoder.

Fig. 6.3   Functional diagram of a 4-bit random access memory.

Known as 'RAM'.

With this logical structure, any storage location can be accessed, depending on the value of the address. The memory locations may be read and written to in any sequence without having to be scanned; hence the name 'random access memory'.

Memory chips are available in sizes up to 64M-bits which require 26 address lines. They can also be byte organized, where a given address will access more than 1 bit in parallel.

An $n$-bit byte-organized memory uses $n$ memory planes operating off a common decoder, as shown in Fig. 6.4. The data input and output channels are now $n$ bits wide. Commercially available memories are discussed further in Chapter 8.

A memory can act as a logic function simply by storing a truth table, and as any truth table can be loaded into store, the device is both programmable and universal. A bit-organized memory only has one data input line, irrespective of its size, through which each and every storage location can be entered. This contrasts with the multiplexer which requires $2^n$ data inputs for $n$ address lines. A memory is therefore capable of implementing large logic systems and is an attractive alternative to multilayer multiplexer systems.

The memory is initially cleared – all locations are reset to 0.

A memory is programmed to perform a combinational logic function by first obtaining the complete truth table of the function. The logic variables will determine the number of address lines and hence the size of the memory required. Each address line represents one input variable. For each input state in the truth table that requires an output at 1, the inputs are applied to the address lines and the location accessed is set to 1. The memory, in read mode, will then perform

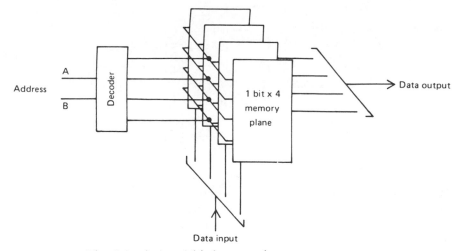

Fig. 6.4   A 4 × 4-bit byte random access memory.

the logic function specified by the truth table. Each input state will address a memory location whose contents correspond with the output condition on the truth table.

When using memories as logic functions, there is no real need, or advantage, to minimize the logic function beyond the elimination of any redundant variables. The size of the memory is determined by the number of dependent variables in the function, and any minimization that does not result in the total elimination of one or more variables from the function will not reduce the storage requirements.

$F = ABC + AB\overline{C}$ requires an 8-bit memory, but if simplified to $F = AB$, a 4-bit memory will suffice.

An $n$-variabled function requires $2^n$ bit memory.

---

Design a combinational logic system based on a memory, which will indicate when a 3-bit binary number **X** is in the range $3_{10} \leqslant \mathbf{X} \leqslant 5_{10}$.

**Worked Example 6.3**

*Solution*   The truth table for the range detector $F = f(X_2 X_1 X_0)$ where $\mathbf{X} = X_2 X_1 X_0$ is

| $X_2$ | $X_1$ | $X_0$ | $F$ |
|-------|-------|-------|-----|
| 0 | 0 | 0 | 0 |
| 0 | 0 | 1 | 0 |
| 0 | 1 | 0 | 0 |
| 0 | 1 | 1 | 1 |
| 1 | 0 | 0 | 1 |
| 1 | 0 | 1 | 1 |
| 1 | 1 | 0 | 0 |
| 1 | 1 | 1 | 0 |

F is dependent on all three input variables $X_2$, $X_1$ and $X_0$; therefore an 8-bit memory is needed. The truth table may be loaded, by first clearing the memory and then writing logical 1 in addresses 011, 100 and 101. The address lines are driven by the input variables and when in the read mode, the memory output gives the desired functions. The memory configuration is

This may be checked on a K-map.

Byte-organized memories may be set to implement multi-output functions. For example, in a BCD to Gray code converter each input in BCD would address a 4-bit location where the equivalent Gray value is stored.

Furthermore, the address field may be partitioned into two parts, to enable more than one function to be programmed into a large memory. Consider a typical commercial memory of 1k-bits. It will have 10 address lines and can be used as a 10-variabled function. Suppose, however, we have a number of four-variabled functions to perform. The address field can be divided into two. Four lines are required for the input data variables, leaving six lines to be used as the function addresses, as shown in Fig. 6.5.

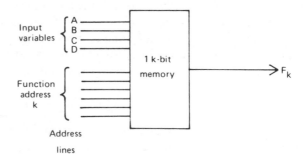

Fig. 6.5   A 1k-bit memory partitioned to perform 64 functions of four inputs.

The same data input lines are used for every function. The memory address is made up of the data inputs and the function address.

The first function $F_0 = f(ABCD)$ is loaded into memory by setting the function address to 000 000 and entering the minterms via ABCD. A second function $F_1 = f(ABCD)$ can be loaded into a different section of memory by setting the function address to 000 001. The minterms of every function are entered via ABCD. Thus a total of $2^6$ or 64 different four-variabled functions can be loaded into the memory.

In use, the desired function is selected via the function address. The input data is applied to ABCD and the memory output gives the function required. Different functions may be selected by changing the function address, although it must be held constant when any given function is being accessed.

**Exercise 6.2**   Design a memory-based logic system that will enable one decade of 8421 BCD to be converted into any of the following codes: 4221, 7421, 5211, $74 - 2 - 1$, $84 - 2 - 1$, $-3642$ BCD codes, a random BCD code of your choice and Gray code (see Table 1.2 for weighted BCD codes). Only one code conversion is required at any given time, but the user must be able to select any specific code.

## Sequential logic using memory elements

A memory circuit can be programmed to perform a sequential logic function provided feedback is incorporated into the system. The address field is partitioned into the external data inputs and the internal states, which are fed back from the output. Figure 6.6 shows a 256 × 4-bit byte-organized memory, configured as a sequential circuit. The memory is programmed via the address and data-in terminals, by compiling an address/contents table from the present/next state and output specifications of the sequential system.

Fig. 6.6  A 256 × 4-bit memory organized as a sequential circuit.

The DFFs in the feedback loop are optional. If these are removed, the circuit becomes asynchronous. The operating time will then be dependent on the access time of the memory, but the system will be susceptible to hazards due to unequal propagation delays in the memory chip. The inclusion of the DFFs synchronizes the feedback and allows the system to switch at a rate determined by the clock driving the flip-flop.

---

Program into a memory system the sequential logic circuit (Worked Example 5.6) which detects the sequence 101 in a stream of bit-serial data.

**Worked Example 6.4**

*Solution*   In Worked Example 5.56 we identified four internal states and obtained the present/next state and output tables. They are

129

where I is the external input, Z the external output, and A and B the state variables.

The address to the memory must comprise the present state (2 bits) and the external state (1 bit). Hence a 3-bit address field is required. Similarly, the output from the memory provides the logic system's output Z (1 bit) and the next states A' and B' (2 bits). The memory needs to be organized in 3-bit bytes giving a minimum requirement of $8 \times 3$ bits. Let the 3-bit address field be IAB and the 3-bit output be ZA'B'. Address 000 represents I = 0 and AB = 00 and we can see from the sequential system specification tables that under these input conditions the output Z needs to be 0 and the next state A'B' is 00. The 3-bit byte addressed by 000 must therefore be set to 000. By referring to the system tables, the required memory contents for each address can be evaluated.

A'B' is the next state, whereas AB is the present state.

They are

| Address | | | Memory contents | | |
|---|---|---|---|---|---|
| I | A | B | Z | A' | B' |
| 0 | 0 | 0 | 0 | 0 | 0 |
| 0 | 0 | 1 | 0 | 1 | 1 |
| 0 | 1 | 0 | 0 | 0 | 0 |
| 0 | 1 | 1 | 0 | 0 | 0 |
| 1 | 0 | 0 | 0 | 0 | 1 |
| 1 | 0 | 1 | 0 | 0 | 1 |
| 1 | 1 | 0 | 0 | 0 | 1 |
| 1 | 1 | 1 | 1 | 1 | 0 |

and the block diagram for the memory is

*Programmable logic arrays*

A programmable logic array (PLA) can be regarded as a memory circuit with a programmable decoder. In a conventional memory the decoder is universal. This means that the $n$ address lines can access each and every one of the $2^n$ storage locations. However, in a PLA, the decoder addresses far fewer than $2^n$ and, under certain circumstances, this may have significant advantages. Consider the combinational function

$$F = ABCDEF + \overline{A}\,\overline{B}\,\overline{C}\,\overline{D}\,\overline{E}\,\overline{F} \qquad (6.4)$$

A 64-bit memory, having six address lines, would be required for this function. If, however, the decoder could be programmed so that

$$X = ABCDEF$$
$$Y = \overline{A}\,\overline{B}\,\overline{C}\,\overline{D}\,\overline{E}\,\overline{F}$$

and then X and Y used to address the memory, only 4 bits would be required and any input other than the two minterms in Equation 6.4 would address location $XY = 00$ and output 0. In a conventional memory these terms would address any one of 62 different cells, all of which must contain 0.

PLAs represent a more efficient way of implementing logic functions and therefore require lower power consumption, fewer and smaller chips for a given system and lower interconnection costs. The electrical behaviour of a PLA is very different from a conventional memory; PLAs do not contain any storage elements as such. They can be regarded as an X.Y matrix of wires and are programmed by making electrical connections between the X and Y intersections. The matrix is divided into two parts – the AND array (equivalent to the decoder) and the OR array (the store). Data is input via the Y wires into the AND array. There is a wire for each input variable and its inverse. The decoder is programmed by burning connections through to the X wires, which represent the implicants of the logic function, and address the OR array. A programmed logic array is shown in Fig. 6.7. The connections are represented by a heavy dot. The Y wires leaving the OR array represent the outputs of the logic functions.

An important electrical property of the array is that the connections to the

If a connection is not required, the internal diode at the node is blown. A PLA is a microcircuit and internal connections cannot therefore be made manually.

Implicants are obtained by combining minterms.

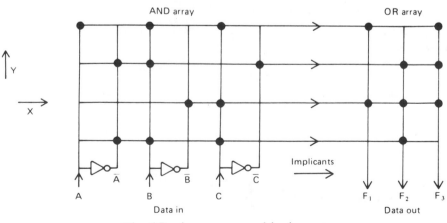

Fig. 6.7   A programmed logic array.

horizontal X wires in the AND array perform the AND function, whereas the connections to the vertical wires in the OR array perform the OR function. The reader should verify that the functions performed by the PLA in Fig. 6.7 are

$$F_1 = ABC + \overline{B}C$$
$$F_2 = \overline{A}B\overline{C} + \overline{B}C + \overline{A}BC$$
$$F_3 = ABC + \overline{A}B\overline{C} + \overline{B}C$$

Use K-maps, Quine–McCluskey algorithm or Boolean algebra. Total minimization may not be needed. The system has to be sufficiently minimal so that the total number of implicants does not exceed the number of X lines.

When using a PLA as a logic system, it must have adequate input and output lines. The third parameter is the number of lines connecting the AND and OR arrays. This represents the total number of different implicants which can occur in the system. A logic system must therefore be minimized so that its set of implicants can be accommodated within the PLA. Common implicants between different output functions within a logic system need only be programmed once into the AND array. The PLA in Fig. 6.7 is a $3 \times 4 \times 3$ array having three inputs, four implicant lines and three outputs.

**Worked Example 6.5**

Program the following system in a $3 \times 3 \times 2$ PLA:

$$F_1 = \Sigma(0,4,5)$$
$$F_2 = \Sigma(0,1,3,4,5)$$

*Solution* The system must be reduced to not more than three implicants. These can be identified on the K-maps of $F_1$ and $F_2$.

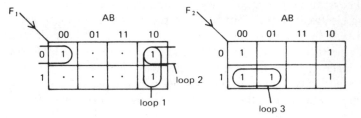

$F_2$ could be minimized further to give $\overline{B} + \overline{A}C$ but this would increase the total number of implicants for the complete system $F_1$ and $F_2$ to four.

By inspection we can see that loops 1 and 2 are common to $F_1$ and $F_2$; the remaining minterms on $F_2$ are covered by loop 3 giving a total of three implicants where

$$F_1 = A\overline{B} + \overline{B}\,\overline{C} \quad \text{and} \quad F_2 = A\overline{B} + \overline{B}\,\overline{C} + \overline{A}C$$

The system can now be programmed into a $3 \times 3 \times 2$ PLA. The connections are as follows:

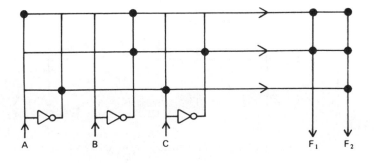

### Programmable arrays of logic (PALs)

Programmable arrays of logic are in many ways similar to the PLAs described in the previous section. They were introduced as a programmable replacement for discrete TTL circuits, and implement the first canonical form of logic. A PAL has a programmable AND array but a fixed OR array, where each output is a function of a prewired number of product terms. The AND array is not universal and the number of product terms that can be set up for an $n$-input PAL is considerably less than $2^n$. This leads to more efficient use of the chip compared with PROM-based systems, where the address decoder must have $2^n$ outputs for an $n$-bit address field. The basic structure of a PAL is shown in Fig. 6.8. This device has three inputs, four product terms and two outputs. Current commercially available PALs can have up to 16 inputs and eight outputs.

Fig. 6.8   A $3 \times 4 \times 2$ PAL.

---

Given the PAL of Fig. 6.8 program the device to convert 3-bit binary (ABC) to 3-bit Gray (XYZ).     **Worked Example 6.6**

*Solution*   The design of the logic to convert binary to Gray code was discussed in Chapter 3 (see Problem 3.7).
   The functions required are

$$X = A$$
$$Y = A\overline{B} + \overline{A}B$$
$$Z = B\overline{C} + \overline{B}C$$

The PAL must generate Y and Z (no logic is required for X). Each PAL output is the OR of two product lines. For the first output of the PAL to be the function Y its associated product terms must be $\overline{A}B$ and $A\overline{B}$. Similarly for output 2 to be function Z, its AND terms must be set to $B\overline{C}$ and $\overline{B}C$.

The programmed array is therefore

PALs can be programmed either by the user or during the last stage of manufacture using a metallic layer to form the desired connections which are then permanent and cannot be erased. These non-volatile devices are sometimes called hard arrays of logic or HALs. The relationship between HALs and PALs is the same as that between ROMs and PROMs. The former are factory programmable and suitable for mass production, whereas the latter are field programmable by the user and ideal for prototyping and small batch production.

The programming characteristics of PALs and ROMs are contrasted below.

| Programmer | | Programmable components | |
|---|---|---|---|
| User | Manufacturer | AND array | OR array |
| FPLA | PLA | Yes | Yes |
| PROM | ROM | No | Yes |
| PAL | HAL | Yes | No |

*The microprocessor as a programmable logic device*

The digital computer is a product of digital electronic engineering and is now, through developments in solid state technology, available as a single-chip microprocessor costing a few dollars. The microprocessor itself can now be regarded as a component in a digital system. It is extremely flexible and can be programmed to perform complex logical and arithmetical operations. Its principal disadvantage is, however, its operating speed. Microprocessors are single-instruction, single-data (SISD) devices having an instruction time of the order of 1 microsecond. To perform a complex operation, a program that may run to several hundred operations is needed. As the instructions are carried out serially – one at a time – the execution time may be long, perhaps of the order of fractions of a second. These times are incompatible with very fast parallel logic systems that can operate in a few microseconds. A microprocessor can therefore only be regarded as a viable component of a larger digital system if the desired operations and response times are relatively slow.

SISD – At any instant an SISD machine can only carry out a single operation on one data word.

For a fuller discussion on the microprocessor, refer to Downton, A. *Computers and Microprocessors* (3rd edn, ITCP, 1992).

## Register transfers

In the designing of digital systems at the circuit level, truth and state tables can be used to specify the behaviour of the system, and we have already developed methods of generating circuits from these specifications. Developments in solid state technology in recent years have, however, made available increasingly complex and cheap integrated circuits, such as adders, counters, timers, memories, encoders and correlators. These and other devices are now becoming the basic components of a digital system.

In order to design a system from complex components, as opposed to gates and flip-flops, we require a means of specifying its behaviour that can then be transformed into interconnections between the components. Register transfer language may be used to represent the order and flow of information between complex components. A register in this context is any self-contained logic system that performs a well-defined function on data. The term is not restricted to shift registers. Adders, correlators, timers, counters, etc., can all be regarded as registers.

A typical register transfer statement is

$$T_i: X \rightarrow A; Y \rightarrow B$$

$T_i$ is the control data and determines when the operation takes place. It is separated from the rest of the statement by a colon (:). 'X $\rightarrow$ A' reads 'input X is transferred to store A'. Simultaneous operations at time $T_i$ are separated by a semi-colon (;).

A register transfer statement can define an operation between several registers.

$$T_j: (A) + (B) \rightarrow A$$

which reads: at time $T_j$ the contents of stores A and B are added together and the sum returned to store A.

To enable the sequence of operations to be carried out in the correct order, a control circuit is required to produce signals to initiate the operations in the order specified by $T_i$ (see Fig. 6.9a). A master clock synchronizes the control circuit and a control unit generates the desired sequence of signals. In complex systems, the controller may be dependent on the result of operations in the digital system (see Fig. 6.9b).

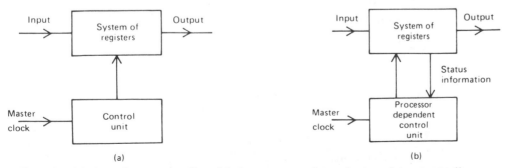

Fig. 6.9  (a) A register controller. (b) A processor dependent register controller.

135

**Worked Example 6.7**  Compile a register transfer specification and design a system to generate two's complement of an input, using an adder and a store.

*Solution*  Two's complement is obtained by inverting the input and adding 1 to the least significant place. We have an adder and a store available, and a register transfer specification for this problem is

$T_1: 0 \rightarrow A$ \qquad \clear store

$T_2: X \rightarrow A$ \qquad \input number

$T_3: \overline{A} + 1 \rightarrow A$ \qquad \generate two's complement

$T_4: A \rightarrow Output$ \qquad \output result

Operation $T_1$ can be achieved by using the clear control on the store. At time $T_2$ the input is applied to the circuit and the store clocked. $T_3$ requires the inverse output of the store to be gated into the adder. The carry-in must be set to 1 and the output from the adder returned to the store. Finally, at time $T_4$ the store contents are gated into the output lines.

The following structure satisfies these requirements:

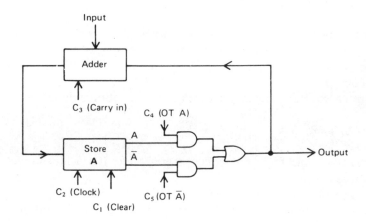

The binary lines represent a parallel word and gating logic is needed in order to output either $A$ or $\overline{A}$ from the store. The five control lines are

$C_1$ – Clear store

$C_2$ – Clock store

$C_3$ – Set adder carry-in to 1

$C_4$ – Output $A$ from store

$C_5$ – Output $\overline{A}$ from store

To generate two's complement, the input word is applied to the adder and the control lines set to 1 in the following order

Assume store is edge triggered... $C_2$ can therefore be applied at the same time as $C_3$ and $C_4$.

$T_1: C_1 = 1$

$T_2: C_2 = 1$

$T_3: C_4 = 1, C_3 = 1, C_2 = 1$

$T_4: C_5 = 1$

The controller can take the form of a 2-bit pure binary counter, driven from a master clock. If the output bits of the counter are $B_1 B_0$ then

Control lines are 0 at all other times.

$$C_1 = \overline{B}_1 \overline{B}_0$$
$$C_2 = \overline{B}_1 B_0 + B_1 \overline{B}_0 = B_1 \oplus B_0$$
$$C_3 = C_4 = B_1 \overline{B}_0$$
$$C_5 = B_1 B_0$$

The controller circuit is therefore

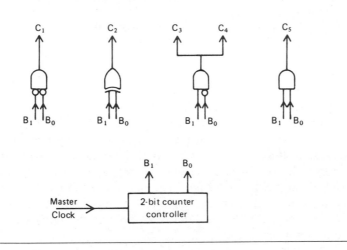

---

Design an arithmetic unit together with a controller that is capable of adding or subtracting two numbers. The arithmetic unit should be based on a parallel full adder and storage registers.

**Exercise 6.3**

### The interconnection problem

When building a digital system from registers, the number of interconnections between modules soon becomes a significant, if not insuperable, problem. Take a system of four 16-bit stores **A**, **B**, **C** and **D** where data transfer between any two stores is permissible. A total of 12 data highways as shown in Fig. 6.10 are needed, each containing 16 wires, if the system is to operate in parallel, giving a grand

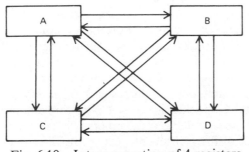

Fig. 6.10   Interconnection of 4 registers.

total of 192 wires. This is in addition to any gating logic. Every input line requires an OR gate to enable inputs from any other register and each output line requires an AND gate so that its output can be selected by a control signal. As more registers are introduced into this system, the number of interconnections increases according to the square law

$$C = (N^2 - N)K \tag{6.5}$$

where $C$ is the number of interconnections, $N$ the registers and $K$ the word length. Hence for a moderately complex system of 16 registers operating on 16-bit words, a conventional interconnection scheme would require 3840 wired links. In most cases this would be regarded as unacceptable. The interconnection problem can be dramatically reduced by using tristate drivers and a data bus.

### The tristate gate

Some tristate gates are disabled with a logical 1 on the control input. Check with manufacturer's data sheet.

A tristate gate, as its name implies, has three output states. The gate also has an additional control input. When the control input is at logical 1 the gate behaves as an ordinary logic gate. A logical 0 on the control disables the gate and the output enters its third state, having a high impedance between the earth and power supply rails. In this state, the gate can only supply or sink a few microamps of current and can neither drive nor load any device connected to it. The gate is effectively disconnected from the circuit. Common tristate gates include the NAND, NOR, INVERTER and DRIVER. Tristate flip-flops are also available. The symbols are given in Fig. 6.11.

A driver is a single input (A), single output (F) gate where $F = A$. It has a high power output and can be used to drive a large number of gates. It is sometimes called a BUFFER. Do not confuse with a buffer store, which simply holds information until required.

Fig. 6.11 Tristate gates. (a) NAND. (b) NOR. (c) INVERTER. (d) DRIVER. (e) JKFF.

### The bus concept

The interconnection problem may be greatly reduced in a register-based digital system by using tristate gates and a bus. A bus is a data highway that runs throughout a digital system. Connections are made from the bus to the inputs of all the registers in the system, and the outputs are connected via tristate drivers back onto the bus. Figure 6.12 shows the four-register problem organized on a bus structure.

Registers are often described as being 'hung on the bus'.

When all the control inputs to the tristate drivers are at 0, there is no data on the bus. In order to carry out a data transfer from, say, registers **C** to **A**, the control $E_c$ to register **C**'s tristate is set to 1. The contents of that register will be output onto the bus and available at the inputs of all the registers. A pulse is then applied to the clock of the register into which the data is to be written (in this case **A**) and the transfer from **C** to **A** is completed.

138

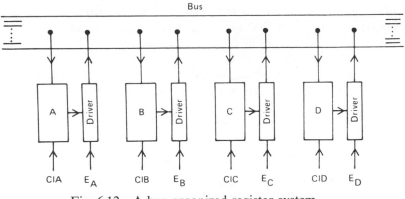

Fig. 6.12   A bus-organized register system.

Bus systems can be readily extended. Additional registers and drivers may simply be connected to the bus and existing registers do not have to be modified in any way. Almost all large digital systems and computers operate on a bus structure and may have more than one bus if high speed operation is required.

### Fault diagnosis

An integrated circuit may contain tens of thousands of electronic elements, each of which can develop faults. It is also very difficult to recognize most faults in a digital system. If a fault develops in an analogue circuit, for example a television set, it will result in distortion and perhaps loss of picture or sound. It is generally quite obvious that something is wrong. However, a failure in a digital system is likely to result in a single bit on an output having the wrong value, a zero instead of a one or vice versa. The output will be a valid binary number but it will have the wrong value. It is therefore essential to test digital systems during their operational life for faults that cannot be observed in the output data.

Testing involves applying a set of inputs to the circuit and knowing the outputs that should be generated. The actual outputs are compared with the expected outputs and if they differ a fault is present. If we examine a combinational circuit having $n$ inputs, there are a total of $2^n$ possible input states each of which has a well-defined output. The system must be tested on every input state in order to ascertain if any faults are present. This exhaustive testing amounts to checking the circuit against its truth table and is viable provided $n$ is small. In larger systems exhaustive testing will take up too much time.

Consider a 64k-bit memory. It has 16 address lines and 65 536 storage locations. A total of 131 072 tests are required to check that a one or a zero can be stored in each location. Unfortunately, however, this may not reveal all the faults in the circuit. The fault could be dependent on the contents of the memory and only occur when a particular pair of locations are set to 1. There are $^{65536}C_2$ or approximately $2 \times 10^9$ different pairs of locations which can be set to 1. Scanning the memory for each store setting, assuming 1 microsecond per read operation, would take

$$65\,536 \times 2 \times 10^9 \times 10^{-6} \text{ seconds}$$

or nearly 100 years. Clearly exhaustive testing is impractical except for very simple systems.

1k $= 2^{10} = 1024$ bits; therefore 64k $= 65\,536$ bits.

In practical fault detection, certain assumptions are made. Only 'stuck' faults are considered. These are the most commonly occurring faults, owing to the elements in the circuit becoming open- or short-circuited and result in data lines being stuck at either logical 0 or 1.

Furthermore, only one fault is assumed to occur at a given time, and diagnosis is assumed to be sufficiently frequent to enable a fault to be rectified before the next one occurs.

The combinational circuit that generates the carry-in a single-bit full adder is shown in Fig. 6.13. Its equation is

$$F = AB + AC + BC \tag{6.6}$$

Stuck faults can occur on the data lines a to j, and the aim of a fault-assessment procedure is to determine a set of non-exhaustive tests that will reveal 'stuck' faults on these lines.

Suppose a fault causes g to be stuck at 0. The function will become

$$F(g) = AC + BC \tag{6.7}$$

In order to detect this fault we require an input that will either give $F = 1$ and $F(g) = 0$ or $F = 0$ and $F(g) = 1$. If the test data produces the same output from the faulty and fault-free circuits, then the fault will not be detected. Therefore the EXCLUSIVE OR of F and F(g) must have a value of 1:

$$F \oplus F(g) = 1 \tag{6.8}$$

By substituting for F and F(g) in Equation 6.8, a suitable test can be obtained:

$$(AB + AC + BC) \oplus (AC + BC) = 1$$

gives

$$AB\overline{C} = 1 \tag{6.9}$$

The fault can therefore be detected by ABC = (110). This input when applied to the fault-free circuit gives an output of 1, but if line g is stuck at 0, the output becomes 0.

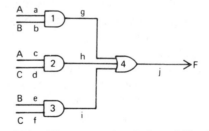

Fig. 6.13   The carry circuit in a full adder.

**Exercise 6.4**   In the following circuit, the output of gate 3 is stuck at 1. Devise a test to identify this fault.

140

*Indistinguishable and undetectable faults*

If a circuit is in its minimal form, all individual 'stuck' faults can be detected. Unfortunately, there is not necessarily a unique test for every fault. Different faults may be detected by the same test and it may be impossible to distinguish between them.

Consider the carry generator shown in Fig. 6.13. Suppose input line c is stuck at 1. The equation of the faulty circuit becomes

$$F(c) = AB + C + BC \qquad (6.10)$$

The test that detects this fault is given by

$$F \oplus F(c) = 1$$

$$\therefore (AB + AC + BC) \oplus (AB + C + BC) = 1 \qquad (6.11)$$

Equation 6.11 need not be evaluated using algebra. Compare K-maps to see which input values set F to 1 and $F_c$ to 0 or vice versa.

giving

$$\overline{A}\,\overline{B}C = 1$$

The fault will therefore be detected by input state

$$ABC = (001)$$

Now, if line e is stuck at 1, the equation of the circuit containing the new fault becomes

$$F_e = AB + AC + C$$

and

$$F_e \oplus F = 1$$

also gives

$$ABC = (001)$$

Hence e stuck at 1 requires the same test as c stuck at 1. ABC = 001 will detect at least two different faults.

In order to determine whether a test exists that will distinguish between the two faults, the EX.OR of the functions of the two faulty circuits is calculated. If it is zero for all inputs, then the faults are indistinguishable.

So

$$F_e \oplus F_c = (AB + C + BC) \oplus AB + AC + C \qquad (6.12)$$

By EX.ORing corresponding cells on the K-maps of $F_e$ and $F_c$ we obtain

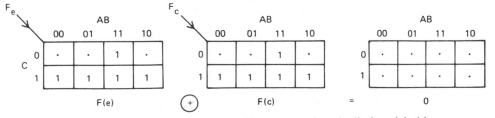

$F_e \oplus F_c$ is zero for all inputs, and the faults are therefore indistinguishable.

In non-minimal circuits, it may be impossible to detect some faults. Therefore if automatic fault detection is envisaged, a system must be designed in its minimal form.

**Worked Example 6.8**

Show that it is impossible to detect a 'stuck-at-0' fault on the output of gate 3 in the following circuit:

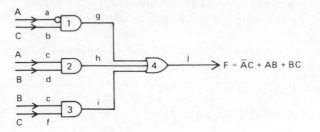

$$F = \overline{A}C + AB + BC$$

Evaluate truth tables for F and F(i) – they are identical.

*Solution*   The equation of the faulty circuit is

$$F(i) = \overline{A}C + AB + 0 = \overline{A}C + AB$$

A test to detect i 'stuck at 0' must satisfy

$$F \oplus F(i) = 1$$

But

$$(\overline{A}C + AB + BC) \oplus (\overline{A}C + AB) = 0$$

$\therefore$ No test will detect fault i 'stuck at 0'

---

**The fault matrix**

In the previous section, a strategy for locating individual faults was given. The basic principles can be extended in order to identify all faults that are detectable within a circuit, together with their test patterns. This results in the fault matrix which is an array relating test patterns to the faults they can detect. To obtain a fault matrix a list of all possible single 'stuck' faults is obtained, and the resulting outputs from the faulty circuits calculated.

The output is 1 if inputs A and B are the same value. Another name is EXCLUSIVE NOR.

Using the 'equivalence' function as an example the fault-free circuit has the equation

$$F = AB + \overline{A}\,\overline{B}$$

The NAND version of the circuit is

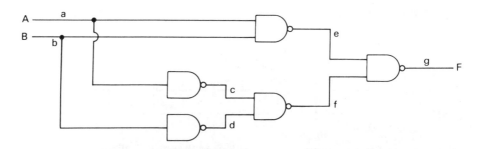

Each fault can be either s-a-1 or s-a-0.

'Stuck' faults can occur on lines a to g giving a total of 14 different single faults. The functional behaviour of all the faulty circuits is given in Table 6.1.

**Table 6.1  Table of 14 fault functions in an 'equivalence' circuit**

| | | Fault free | | | | | | | | Outputs — Faults | | | | | | |
|---|---|---|---|---|---|---|---|---|---|---|---|---|---|---|---|---|
| | | | a–0 | a–1 | b–0 | b–1 | c–0 | c–1 | d–0 | d–1 | e–0 | e–1 | f–0 | f–1 | g–0 | g–1 |
| A | B | F | $F_{a0}$ | $F_{a1}$ | $F_{b0}$ | $F_{b1}$ | $F_{c0}$ | $F_{c1}$ | $F_{d0}$ | $F_{d1}$ | $F_{e0}$ | $F_{e1}$ | $F_{f0}$ | $F_{f1}$ | $F_{g0}$ | $F_{g1}$ |
| 0 | 0 | 1 | 1 | 0 | 1 | 0 | 0 | 1 | 0 | 1 | 1 | 1 | 1 | 0 | 0 | 1 |
| 0 | 1 | 0 | 0 | 1 | 1 | 0 | 0 | 0 | 0 | 1 | 1 | 0 | 1 | 0 | 0 | 1 |
| 1 | 0 | 0 | 1 | 0 | 0 | 1 | 0 | 1 | 0 | 0 | 1 | 0 | 1 | 0 | 0 | 1 |
| 1 | 1 | 1 | 0 | 1 | 0 | 1 | 1 | 1 | 1 | 1 | 1 | 0 | 1 | 1 | 0 | 1 |

**Table 6.2  Fault detection table. In each column a one indicates that the test pattern is able to detect the fault**

| Test pattern | | Faulty circuit | | | | | | | | | | | | | |
|---|---|---|---|---|---|---|---|---|---|---|---|---|---|---|---|
| A | B | $F_{a0}$ | $F_{a1}$ | $F_{b0}$ | $F_{b1}$ | $F_{c0}$ | $F_{c1}$ | $F_{d0}$ | $F_{d1}$ | $F_{e0}$ | $F_{e1}$ | $F_{f0}$ | $F_{f1}$ | $F_{g0}$ | $F_{g1}$ |
| 0 | 0 | . | 1 | . | 1 | 1 | . | 1 | . | . | . | . | 1 | 1 | . |
| 0 | 1 | . | 1 | 1 | . | . | . | . | 1 | 1 | . | 1 | . | . | 1 |
| 1 | 0 | 1 | . | . | 1 | . | 1 | . | . | 1 | . | 1 | . | . | 1 |
| 1 | 1 | 1 | . | 1 | . | . | . | . | . | . | 1 | . | . | 1 | . |

**Table 6.3  Reduced fault table. Two groups of three faults are indistinguishable: $F_{c0}$, $F_{d0}$ and $F_{f1}$ and also $F_{e0}$, $F_{f0}$ and $F_{g1}$**

| Test pattern | | Faulty circuits | | | | | | | | | | | | | |
|---|---|---|---|---|---|---|---|---|---|---|---|---|---|---|---|
| A | B | $F_{a0}$ | $F_{a1}$ | $F_{b0}$ | $F_{b1}$ | $F_{c0}$ | $F_{d0}$ | $F_{f1}$ | $F_{c1}$ | $F_{d1}$ | $F_{e0}$ | $F_{f0}$ | $F_{g1}$ | $F_{e1}$ | $F_{g0}$ |
| 0 | 1 | . | 1 | . | 1 | 1 | . | . | . | . | . | . | . | . | 1 |
| 0 | 1 | . | 1 | 1 | . | . | . | . | . | 1 | 1 | . | . | . | . |
| 1 | 0 | 1 | . | . | 1 | . | . | . | 1 | . | 1 | . | . | . | . |
| 1 | 1 | 1 | . | 1 | . | . | . | . | . | . | . | . | . | 1 | 1 |

A fault detection table can be derived from the fault function table. For every fault which is detectable a one is placed in the table against the appropriate test input on AB. The detectable faults are shown in Table 6.2.

A fault can be detected for those inputs that are a solution to the equation

$$F \oplus F_i = 1$$

where $F_i$ is the function of the faulty circuit.

The fault detection table can now be reduced by combining columns that are identical as shown in Table 6.3.

It is now possible to find the essential tests. The process is similar to the selection of essential prime implicants from a prime implicant table. A test (input AB) pattern is essential if it is the only one that detects a particular fault. If a fault

See Quine–McCluskey minimization algorithm.

can only be detected by one test pattern, the fault column has only one entry and that test is essential. $AB = 00$ is essential for ($F_{c0}$, $F_{d0}$ and $F_{f0}$). It also covers $F_{a1}$ and $F_{b1}$ and $F_{g0}$. $AB = 00$ is essential for $F_{c1}$, and also covers $F_{a0}$ and the set $F_{e0}$, $F_{f0}$ and $F_{g0}$. The procedure is continued until all the faulty circuits have been detected using the minimum number of test patterns. In this example all four test patterns are essential for complete cover. As the logic system becomes large, total cover can usually be obtained with less than $2^n$ input patterns for an $n$-input system, and hence savings in test time can be achieved compared with exhaustive testing schemes.

### Boolean difference method

The Boolean difference method is a more formal technique for locating specific faults. Consider the general function

$$F = f(x_1, x_2, \ldots, x_i, \ldots, x_n)$$

Note $x_i$ has become $\overline{x}_i$ in the equation of the faulty cct.

and an input $x_i$ is faulty. Let us assume that it is the inverse of its correct value so that $x_i$ becomes $\overline{x}_i$, and hence both 'stuck-at-1' and 'stuck-at-0' faults will be taken into account. The faulty function is

$$F_{xi} = f(x_1, x_2, \ldots, \overline{x}_i, \ldots, x_n)$$

The Boolean difference is defined as

$$\frac{dF}{dF_{xi}} = F \oplus F_{xi}$$

and can be evaluated either mathematically or by EX.ORing corresponding cells on the K-maps of $F$ and $F_{xi}$.

Using the equivalence function as an example

$$F = AB + \overline{A}\,\overline{B}$$

and a fault on input A gives:

$$Fa = \overline{A}B + A\overline{B}$$

Evaluate $(AB + \overline{A}\,\overline{B}) \oplus (A\overline{B} + \overline{A}B)$.

The Boolean difference is:

$$\frac{dF}{dFa} = \overline{A}\,\overline{B} + A\overline{B} + \overline{A}B + AB$$

implying that every input pattern is a test for a fault on A. This set tests for both types of fault. Those patterns which test for A s-a-0 must attempt to drive A to 1. They are $A\overline{B}$ and $AB$ or in terms of the input patterns

$$'AB' = 10 \text{ or } 11$$

The s-a-1 tests are $\overline{A}B$ or $\overline{A}\,\overline{B}$, giving $|AB| = 01$ or $00$.

See diagram on page 142.

The Boolean difference can be used to obtain tests for intermediate faults between the inputs and the outputs. Consider a fault on line e in the 'equivalence' circuit:

$$F = AB + \overline{A}\,\overline{B} = \overline{\overline{A}\,\overline{B} . \overline{\overline{A}\,\overline{B}}}$$
$$E = \overline{A}\,\overline{B}$$

where E, the data on line e, is a function of the inputs A and B. By substituting for E in the equation of the circuit we obtain

$$F = \overline{E.\overline{\overline{A}.\overline{B}}}$$

and inverting E we obtain the equation for the faulty circuit:

$$F = \overline{\overline{E}.\overline{\overline{A}.\overline{B}}}$$

Note that e is a reference point in the circuit. E is the logic function present at point e.

The Boolean difference between the fault-free function and the function with line e faulty is given by the equation

$$\frac{dF}{dFe} = F \oplus Fe = A\overline{B} + \overline{A}B + AB$$

hence $|AB| = 10$, 01 or 11, and these are tests for faults on line e. Those that test for e s-a-1 must attempt to drive e to 0.

$E = \overline{A}\,\overline{B}$.

The only pattern that will drive E to 0 is AB = 11. The other patterns AB = 10 or 01 attempt to drive E to 1 and they are tests for e s-a-0. Cross-referencing with the fault matrix, it can be seen that these test patterns are those that detect s-a-0 and s-a-1 on line e.

Boolean difference provides a means of obtaining tests for specific faults whereas the fault matrix approach gives total cover for all testable faults. Both methods can be computerized, but in practice they cannot be applied to large circuits having more than about 100 possible fault lines without incurring large memory and time overheads when generating suitable test patterns.

## Path sensitization

One of the most widely used techniques in fault diagnosis is path sensitization. In its simple one-dimensional form, a fault location is considered and the input data chosen so that only one path from the input to the output via the fault location is activated. Input data is then switched so the path is required to change between 0 and 1. If the circuit is fault free the switching can be observed on the output. If there is a fault the output will remain at either logical 0 or 1.

The principle of path sensitization can be demonstrated using individual gates. Take for example an AND gate:

If a fault is suspected on the output, a single path, say B to F, can be sensitized by setting A to 1. If B is then switched 0, 1, 0, etc., this waveform should appear at the output F, provided no faults are present. The technique can be applied to more complicated circuits. Take for example the carry circuit from a full adder:

Consider a potential fault on line h. We require g and i to be 0 so that h is propagated to the output F. $g = 0$ requires either $A = 0$ or $B = 0$, and $i = 0$ requires either $B = 0$ or $C = 0$. For h to be activated either A is 1 and C has to be switched 1, 0, 1 or C held at 1 and A switched. The overall input status to satisfy these three requirements is $B = 0$ and $C = 1$. The path A–c–h–j–F is then activated, and if fault free then $F = A$, and the switching pattern on A will be reproduced at the output F. Any fault on h (or elsewhere on the path) will cause F to remain constant. The one-dimensional path sensitization method has limitations. It may not be possible to sensitize in turn all paths through the circuit, and hence complete fault cover cannot always be achieved. The method has been developed to allow multiple path sensitization and is becoming one of the most widely used techniques for locating faults in logic systems.

**Fault detection in sequential circuits**

The testing of sequential circuits is far more difficult than for combinational systems, and is currently the subject of much research. Problems arise due to the output being a function not only of the input variables but also of the internal state variables which are inaccessible and cannot be controlled directly from the input. Techniques which show potential for fault detection in sequential logic involve the use of feedback shift registers which can be used to monitor data at nodes in the circuit. These devices are known as signature analysers.

A 3-bit signature analyser is shown below.

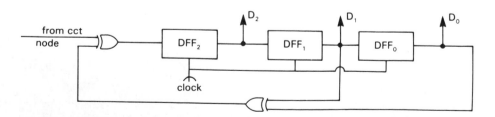

The data from the node of the circuit is EXORed with the feedback line and so the natural cycle of the feedback shift register is disturbed. If the sequential system, after having been initialized, is clocked for a known number of times (say 30) and the circuit is functioning correctly, the output of the feedback shift register $D_2 D_1 D_0$ will be in a definite state. This is known as the signature and it can be determined when the circuit is functioning correctly. If a fault develops, a different sequence of data will occur at the circuit nodes and this will divert the feedback shift register into a different state sequence. The final signature will be changed. This can be detected and indicates that a fault has occurred on the system. The typical length of a signature is in practice of the order of 16 bits and it does provide a facility for monitoring the fault status of a circuit node.

The FSR must always be started from the same initial state (not $00\dots0$).

## Fault-tolerant systems

The ultimate logic system is one that is tolerant to faults and does not malfunction when a fault occurs. This is usually achieved by having massive redundancy. In some of the systems used in space vehicles, as many as five processors are used to perform the same operations. The outputs are correlated and a majority decision made on the results. In this way, up to two systems can fail without corrupting the outputs.

At a less sophisticated level, however, redundancy still provides a tolerance to faults. In Worked Example 6.7, the 'stuck-at-0' fault on line i of gate 3 does not corrupt the output of the circuit. The faulty circuit has the same K-map as the original function, because gate 3 is redundant.

Thus the digital designer is left with a dilemma. Either the designer provides a minimal design that is fault diagnosable and can easily be repaired, or he or she must opt for a highly redundant system that is fault tolerant, but which cannot be fully fault diagnosed and is therefore difficult to service. One would expect the former to have a relatively large number of short periods when non-operational, whereas the latter will give a prolonged period of operation until the number of faults is sufficiently large that they cannot be overcome by the redundancy. An exhaustive service operation would then be required.

## Summary

Programmable logic can perform any desired function without having to make any physical changes to the hardware structure. A multiplexer may operate as a universal combinational logic function. It is programmed directly from a truth table by means of external logic switches on the data-in lines. Multilayer multiplexer logic systems can be devised for larger logic functions. The memory element may also act as a programmable logic function. As the truth table is stored internally, a memory-based logic function does not require the $2^n$ input connections needed for an $n$-variabled function implemented on a multiplexer. The truth table can be entered serially via a single-input line.

A PLA can be regarded as a memory circuit with a programmable decoder. Its structure provides a more efficient means of implementing logic functions in terms of hardware size and power consumption. If memories or multiplexers are used, a function need only be minimized as far as the removal of redundant input variables, whereas with PLAs the system must be reduced to the number of implicants the PLA is capable of accommodating. A PAL offers a programmable AND array but it has a fixed OR array.

The incorporation of feedback within a programmable logic structure enables a sequential logic system to be implemented.

Register transfer language is used to specify the operation of a system based on large scale integrated circuit components (the registers). As the number of registers in a system increases, the number of interconnections becomes unacceptably large. The interconnection problem can be greatly reduced by using tristate gates and a bus structure.

Fault diagnosis of large and complex systems is becoming an increasingly important subject. Test pattern generation based on EXCLUSIVE OR behaviour between faulty and fault-free systems has been examined and finally the possibility of fault-tolerant systems was proposed, by allowing redundancy in the system.

The fault matrix summarizes the extent to which a circuit is fault testable. The Boolean difference technique offers a formal method for devising individual tests. Path sensitization is a practical technique for testing specific faults and signature verification offers some potential for sequential circuit diagnostics.

## Problems

6.1 Determine the input settings required on an 8 to 1 multiplexer if it is to implement the following three-variabled functions:

   (i) $F = \Sigma(2,4,5,7)$
  (ii) $F = \Pi(2,5,6)$
 (iii) $F = \overline{AB + \overline{A(B + C)}}$

6.2 Determine the input settings required on an 8 to 1 multiplier if it is to implement the following functions of four variables:

   (i) $F = \Sigma(1,3,5,6,9,12,14)$
  (ii) $F = \Pi(0,1,2,3,4,5,6,7,12,13)$

6.3 Design a two-layer system of 4 to 1 multiplexers to implement the function

$$F = ABC + \overline{A}\,\overline{B}D + D\overline{E} + AE$$

6.4 An electronic voting system has four voting stations where votes for, against or abstensions can be registered. Devise a suitable logic system using a memory implementation to indicate when a motion is accepted.

6.5 Design a feedback memory circuit to measure, in Gray code, the duration of a pulse, to 5-bits resolution.

6.6 What size PLA is required to implement the following three-function system?

$$F_1 = A\overline{B} + \overline{A}$$
$$F_2 = \overline{A}\overline{D} + ABC$$
$$F_3 = A\overline{C}D + \overline{A}\overline{B}\overline{D}$$

Specify the connections in the array.

6.7 Implement the system

$$F_1 = \overline{A}BC + AB$$
$$F_2(A + B)(A + C)$$
$$F_3 = BC + A\overline{B}$$

in a $3 \times 3 \times 2$ PLA.

6.8 Obtain a register transfer specification of a hardware multiplier which uses a successive addition procedure.

6.9 Examine the following circuit and determine which tests, if any, can identify a stuck-at-1 fault on any of the lines a to h.

6.10 Repeat Problem 6.9 for a stuck-at-0 fault on any one of the lines a to h.

6.11 Obtain the fault matrix of an EXCLUSIVE OR function, constructed out of NAND gates.

6.12 Use the path sensitization method to test for a fault on line f in Problem 6.9.

# 7 Design of arithmetic processors

**Objectives**

☐ To design logic circuits for arithmetic.
☐ To examine ways of increasing the speed of arithmetic units.
☐ To develop a programmable arithmetic unit capable of performing a number of arithmetic calculations.
☐ To introduce the concept of a programming language.

An important application of logic is in an arithmetic unit. Such a device must be capable of being reconfigured to enable it to carry out a range of numerical operations. This ability to carry out many different operations, depending on the problem needing to be solved, leads to the concept of a programmable system – a computer. The main components of an arithmetic unit will be developed in this chapter and a simple system controller will be specified. An elementary programming language for operating this system will be developed.

## Ripple-through-carry addition

See page 59.

The addition of two binary inputs has already been considered in Chapter 3. The combinational logic designs for a full adder circuit which accepts input bits A and B and generates the sum A plus B, plus any carry-in together with the resulting carry-out, are given in Fig. 3.9. This full adder circuit can only provide the sum of two single-bit inputs – a very limited device. To extend the resolution of the adder, the cells can be cascaded where each cell receives one corresponding bit from each of the two input numbers to be added together and the carry is propagated from cell to cell as shown in Fig. 7.1.

Any system must be designed for the worst case – it could happen in practice.

Such a system is known as a ripple-through-carry adder because in the worst case each cell produces a carry and each successive cell cannot generate the correct summation until the previous, i.e. lesser significant cell, has generated its carry output. Consider a ripple-through-carry adder of 8-bits resolution, performing the addition $A + B$ where

$$A = 127_{10} \quad \text{and} \quad B = 1_{10}$$

Fig. 7.1   A ripple-through-carry adder.

In binary this is

$$01111111_2 + 00000001_2$$

The external inputs to the most significant cell (cell 7) are $A_7 = 0_2$ and $B_7 = 0_2$, which might suggest that the most significant sum bit $S_7$ is 0. However, in adding the two numbers together, the answer is $10000000_2$; the most significant bit $S_7$ is 1. The ripple-through-carry adder will only produce $S_7 = 1$ when the carry of the least significant cell $C_0(\text{out})$ is generated and propagated to cell 1 which will in turn generate $C_1(\text{out}) = 1$; then $C_2(\text{out}) = 1$ from cell 2. For this addition the carry has to be propagated through every cell and the correct sum output does not occur until the carry reaches the most significant cell which then responds with the correct sum output $S_7 = 1$, giving the correct output from the adder:

$$S = S_7 S_6 \ldots S_0 = 10000000_2 = 128_{10}$$

In general $\mathbf{A} = A_7 A_6 \ldots A_0$

Whilst this is the worst case and the carry has to ripple through every stage of the adder, it can occur in the addition of certain pairs of input numbers. The designer must therefore make allowances for this worst case when specifying an arithmetic unit containing ripple-through-carry adders. If the delay in each cell is $2\,\mathrm{d}t$, then the operating time $T$ for a ripple-through-carry adder is

$$T = n \times 2\,\mathrm{d}t \tag{7.1}$$

where $n$ is the number of cells.

This linear increase in operating time can impair performance for high resolution adders, i.e. 32- and 64-bit word length.

If the gate delay $\mathrm{d}t$ is $10^{-7}$ seconds, the operating time of a 32-bit adder will be $6.4\,\mu\mathrm{s}$.

### Serial adder

An alternative structure to the ripple-through-carry adder is the serial adder. Here, only one full adder cell is required regardless of the resolution of the adder, thereby reducing the combinational logic requirements. In the serial adder, a bit by bit addition is performed as the addends are shifted into the full adder from storage registers, starting with the least significant bits. The block diagram of the serial adder is shown in Fig. 7.2.

A single data flip-flop is required to hold the carry bit after each operation of the full adder. The sum bits generated by the addition can be stored in one of the addend registers (register **A** in Fig. 7.2) by clocking the data into the most significant end of the register as the addend is output from the least significant end of the register. After $n$ operations of the full adder the complete sum will then

Fig. 7.2   A serial adder.

be in register **A**. The maximum operating speed of the serial adder will be greater than the worst case time for the ripple-through-carry adder, because in addition to the $n$ operations of the full adder, time must be allowed to transfer data into and out of the registers.

The serial adder can, however, be easily upgraded to a higher resolution simply by increasing the register lengths. No extra addition logic is required.

**Parallel adders**

An adder circuit can be designed as canonical form logic and as such will only have a delay of 2 dt regardless of resolution of the adder. The number of inputs and hence the complexity of the circuit will, however, increase with increasing resolution of the adder.

---

**Worked Example 7.1** Design a 2-bit adder circuit as a 1st canonical form system.

*Solution*  Let the addends be $\mathbf{A} = A_1A_0$ and $\mathbf{B} = B_1B_0$.

The maximum resolution of a 2-bit adder is 3 bits $(S_2S_1S_0)$ as

$11_2$ is the maximum value of a 2-bit input.

$$11_2 + 11_2 = 110_2$$

Now, specify a truth table giving the values of the outputs $S_2S_1S_0$ for all possible values of **A** and **B**.

| $A_1$ | $A_0$ | $B_1$ | $B_2$ | $S_2$ | $S_1$ | $S_0$ |
|---|---|---|---|---|---|---|
| 0 | 0 | 0 | 0 | 0 | 0 | 0 |
| 0 | 0 | 0 | 1 | 0 | 0 | 1 |
| 0 | 0 | 1 | 0 | 0 | 1 | 0 |
| 0 | 0 | 1 | 1 | 0 | 1 | 1 |
| 0 | 1 | 0 | 0 | 0 | 0 | 1 |
| 0 | 1 | 0 | 1 | 0 | 1 | 0 |
| 0 | 1 | 1 | 0 | 0 | 1 | 1 |
| 0 | 1 | 1 | 1 | 1 | 0 | 0 |
| 1 | 0 | 0 | 0 | 0 | 1 | 0 |
| 1 | 0 | 0 | 1 | 0 | 1 | 1 |
| 1 | 0 | 1 | 0 | 1 | 0 | 0 |
| 1 | 0 | 1 | 1 | 1 | 0 | 1 |
| 1 | 1 | 0 | 0 | 0 | 1 | 1 |
| 1 | 1 | 0 | 1 | 1 | 0 | 0 |
| 1 | 1 | 1 | 0 | 1 | 0 | 1 |
| 1 | 1 | 1 | 1 | 1 | 1 | 0 |

Therefore the three canonical circuits in the adder are

$$S_2 = \Sigma(7,10,11,13,14,15)$$
$$S_1 = \Sigma(2,3,5,6,8,9,12,15)$$
$$S_0 = \Sigma(1,3,4,6,9,11,12,14)$$

These functions can be minimized by using either Karnaugh maps or the Quine–McCluskey algorithm.

---

In the 2-bit adder designed in the above worked example, $S_2$, $S_1$ and $S_0$ can operate at the same time, i.e. in parallel. The operating time is therefore only that of a single canonical form circuit. Unfortunately, however, parallel adders have to be designed for a specific resolution and cannot be expanded without major logic redesign. If the parallel adder in Worked Example 7.1 were to be expanded to input a resolution of 3 bits, functions $S_2$, $S_1$ and $S_0$ could be used, but a completely new function $S_3$, which would be dependent on all six input variables $A_2$, $A_1$, $A_0$, $B_2$, $B_1$, $B_0$ needs to be designed. This function is more complex, being dependent on six variables, than the lower significance output bits $S_2$, $S_1$ and $S_0$. If designs for increasingly higher resolution parallel adders are attempted, the complexity, measured by the maximum number of input variables, of the most significant sum bit is $2n$, where $n$ is the resolution of the adder. This limits the practicality of the approach as such functions are increasingly difficult to realize in canonical forms. For example, in a 32-bit parallel adder, the most significant sum output bit is a function of 64 input variables.

*Operating time 2 d$t$.*

### Look-ahead-carry adder

An alternative architecture for fast addition is the look-ahead-carry adder. This differs from the parallel adder as all the carry bits are generated in parallel as soon as the numbers are input. The carry bits are then added to the appropriate input bits using the summing circuit from the full adder. In order to generate the carry bits, the exact conditions under which a carry is generated need to be examined. Consider input bits $A_i$ and $B_i$. If both are zero, no carry bit is ever generated. If $A_i$ and $B_i$ are both 1 then a carry is always generated. Finally if $A_i = 1$ and $B_i = 0$, or $A_i = 0$ and $B_i = 1$, then the carry depends on the carry-out from the previous $(i-1)$th cell.

*$S = A_i \oplus B_i \oplus C_i(\text{in})$. See page 59.*

*$C_{i-1}(\text{out}) = C_i(\text{in})$*

Two new functions can now be defined. The carry generate G is

$$G_i = A_i . B_i \tag{7.2}$$

This is the condition where there is always a carry bit. The carry propagate P is

$$P_i = A_i \oplus B_i \tag{7.3}$$

and identifies the condition where the carry-in to the $i$th cell must be propagated.

The carry-out equation for the $i$th cell can now be defined as

$$C_i(\text{out}) = G_i + P_i C_i(\text{in}) \tag{7.4}$$

This gives a carry-out bit set to 1 if the carry generate condition is true, i.e. $A_i$ and $B_i$ are both 1, or the carry propagate condition is true and there is a carry-in bit set to 1 from the previous cell. By substituting the generate $G_i$ and propagate $P_i$ equations into the carry-out equation $C_i(\text{out})$, a carry generator operating only on the input bits results, and if designed as a canonical form circuit will only introduce a maximum of two gate delays.

*Look-ahead-carry adder is therefore faster than ripple-through-carry, but slower than a parallel adder.*

The carry generate for the least significant bit $C_0(\text{out})$ is

$$C_0(\text{out}) = G_0 + P_0 C_{-1}(\text{in}) \tag{7.5}$$

but as the carry-in to the least significant cell $C_{-1}(\text{in})$ must be 0 then

$$C_0(\text{out}) = G_0 = A_0 B_0 \tag{7.6}$$

In high resolution adders blocks of look-ahead-carry circuits can be used. If so, $C_0(\text{out})$ is a function of $C_{-1(\text{in})}$ and Equation 7.5 should be used.

For the next bit of the adder (bit 1), the carry-out equation $C_1(\text{out})$ is

$$C_1(\text{out}) = G_1 + P_1 C_1(\text{in}) \tag{7.7}$$

but

$$G_1 = A_1 B_1$$
$$P_1 = A_1 \oplus B_1$$

and

$$C_1(\text{in}) = C_0(\text{out}) = A_0 B_0$$

Hence

$$
\begin{aligned}
C_1(\text{out}) &= A_1 B_1 + (A_i \oplus B_1).(A_0 B_0) \\
&= A_1 B_1 + (A_1 \overline{B}_1 + \overline{A}_1 B_1).(A_0 B_0) \\
&= A_1 B_1 + A_1 \overline{B}_1 A_0 B_0 + \overline{A}_1 B_1 A_0 B_0
\end{aligned}
\tag{7.8}
$$

Although $C_1(\text{out})$ is more complex than $C_0(\text{out})$, it is still a two-level canonical form equation depending only on the external inputs $A_i$ and $B_i$ and can operate at the same time as the equation for $C_0(\text{out})$. A similar equation can be developed for $C_2(\text{out})$:

$$C_2(\text{out}) = G_2 + P_2 C_2(\text{in}) \tag{7.9}$$

where

$$G_2 = A_2 . B_2$$
$$P_2 = A_2 \oplus B_2$$

and

$$
\begin{aligned}
C_2(\text{in}) &= C_1(\text{out}) \\
&= A_1 B_1 + A_1 \overline{B}_1 A_0 B_0 + \overline{A}_1 B_1 A_0 B_0
\end{aligned}
$$

By carrying out the above substitutions in Equation (7.9), another two-level circuit results. Although it is a more complex function than $C_1(\text{out})$ or $C_0(\text{out})$, it is still only dependent on the inputs $A_i$ and $B_i$. There is no ripple-through dependency and therefore all the carry calculations can be carried out in parallel and the total calculation time is only $2\,dt$ – the delay of a canonical form logic circuit. The method can be extended to more significant carry bits and in every case a carry can be generated by a canonical form circuit operating only on the input variables. Unfortunately the carry generate functions become increasingly complex as the level of significance increases. Consequently look-ahead-carry generators are sometimes restricted to 4 bits with a ripple-through-carry occurring between the blocks of 4 bits. If this is done, the carry-in to the least significant cell of the block $C_{-1}(\text{in})$ cannot be assumed to be 0 as was the case in Equations

Fig. 7.3   Block diagram of a look-ahead-carry generator adder.

7.5 and 7.6. It is left to the reader to show that Equations 7.6 and 7.9 then become

$$C_0(\text{out}) = A_0B_0 + (A_0 \oplus B_0)C_{-1}(\text{in}) \tag{7.10}$$

$$C_1(\text{out}) = A_1B_1 + (A_1 \oplus B_1)(A_0B_0 + (A_0 \oplus B_0)C_{-1}(\text{in})) \tag{7.11}$$

A block diagram of a look-ahead-carry generator adder is shown in Fig. 7.3.

**Exercise 7.1**

Design a parallel adder and a look-ahead-carry adder, both of 3-bits resolution, and determine which adder technology leads to the most simple implementation in terms of the gate counts using NAND gates in the design. If NAND and EX.OR gates could be used, would this have any effect on the relative economics of either approach?

### The full subtractor

In arithmetic the process of subtracting one number from another has some similarity with addition. A digit by digit approach is used starting with the least significant digits. There is a transfer of information into and out of the cell which is dependent on the values of the external inputs to the cell. These variables are called the borrow-in and borrow-out. The block diagram of the ripple-through-borrow subtractor is shown in Fig. 7.4.

Note the similarity in structure with the ripple-through-carry adder.

Fig. 7.4   A ripple-through-borrow full subtractor.

The full subtractor cell comprises two combinational logic functions, the difference output $D_i$ and the borrow-out $B_i(\text{out})$ and each is a function of three inputs $X_i$, $Y_i$ (the numbers to be input to the subtractor) and $B_{i-1}(\text{in})$, the borrow-in from the previous (less significant) cell. Before specifying the truth table, the inputs $X$ and $Y$ must be defined as either the minuend or the subtrahend. These inputs cannot be interchanged when the subtractor is subsequently used as an arithmetic input. Suppose $X$ is defined as the minuend and $Y$ the subtrahend, such that the subtractor calculates

$$D = X - Y \tag{7.12}$$

By considering all possible binary inputs and resulting outputs to the cell, a truth table (Table 7.1) can be defined as follows:

**Table 7.1  Truth table for a 1-bit full subtractor**

Explanation:

| $X - Y - B(\text{in}) = D$ | | $B(\text{out})$ |
|---|---|---|
| $0 - 0 - 0$ | $= 0$ | $0$ |
| $0 - 0 - 1$ | $= 1$ | $1$ |
| $0 - 1 - 0$ | $= 1$ | $1$ |
| $0 - 1 - 1$ | $= 0$ | $1$ |
| $1 - 0 - 0$ | $= 1$ | $0$ |
| $1 - 0 - 1$ | $= 0$ | $0$ |
| $1 - 1 - 0$ | $= 0$ | $0$ |
| $1 - 1 - 1$ | $= 1$ | $1$ |

| X | Y | B(in) | D | B(out) |
|---|---|---|---|---|
| 0 | 0 | 0 | 0 | 0 |
| 0 | 0 | 1 | 1 | 1 |
| 0 | 1 | 0 | 1 | 1 |
| 0 | 1 | 1 | 0 | 1 |
| 1 | 0 | 0 | 1 | 0 |
| 1 | 0 | 1 | 0 | 0 |
| 1 | 1 | 0 | 0 | 0 |
| 1 | 1 | 1 | 1 | 1 |

From the truth table it can be seen that

$$D = f(XYB(\text{in})) = \Sigma(1,2,4,7) \tag{7.13}$$

and

$$B(\text{out}) = f(XYB(\text{in})) = \Sigma(1,2,3,7) \tag{7.14}$$

Using Karnaugh maps it can be shown that D does not simplify and is

$$D = \overline{X}\,\overline{Y}B(\text{in}) + \overline{X}Y\overline{B(\text{in})} + X\overline{Y}\,\overline{B(\text{in})} + XYB(\text{in}) = X \oplus Y \oplus B(\text{in}) \tag{7.15}$$

The borrow-out function simplifies to

$$B(\text{out}) = \overline{X}Y + \overline{X}B(\text{in}) + YB(\text{in}) \tag{7.16}$$

It can be seen by referring to the equations of a full adder circuit that the difference function D is exactly the same as the sum function of a full adder, but that the borrow-out equation is NOT identical to the carry-out generator. An adder circuit cannot therefore be used for subtraction with unsigned binary-coded numbers.

**Fast subtractors**

The full subtractor is a ripple-through circuit where the worst case operating time is when the borrow-out has to be propagated through every cell, giving an operation time of $2n\,dt$ where $n$ is the number of cells (i.e. the resolution of the subtractor) and $dt$ is the average gate delay. Methods used to speed up adder circuits can also be applied to the subtractor. The overall form of the circuits is identical to that of the adder circuits but the Boolean equations are different. Parallel subtractors can be designed where there is no borrow generation and the difference is calculated directly from the two input numbers. Cellular subtractors, where each cell processes more than 2 bits from each input number, can be specified, and look-ahead-borrow subtractors are also possible.

In practice, however, dedicated hardware subtractors are unusual and not often designed. Subtraction is carried out using adder circuits and in order for this to be possible a modified binary coding has to be used, which indicates whether the number is positive or negative.

Design a parallel subtractor having 2-bits resolution. **Exercise 7.2**

### Signed numbers

In order to indicate the sign of a number, the most significant bit is used to indicate whether the number is positive or negative. The remainder of the number indicates its magnitude. Hence for an 8-bit number

$$\mathbf{X} = x_7 x_6 \ldots x_0 \tag{7.17}$$

$x_7$ is the sign bit and the magnitude is the value of the bits $x_6 \ldots x_0$. It is conventional for the sign bit to be zero for positive numbers, and the largest value of $\mathbf{X}$ is then $01111111_2$ which represents $+127_{10}$. If the number was unsigned the maximum value would be $11111111_2$ or $255_{10}$. Counting down from maximum, with signed numbers, we get $01111110_2$, $01111101_2$, etc. We eventually reach $00000000_2$ which is interpreted as $+0_{10}$. The sign bit can now be set to 1 indicating a negative number and $1000000_2$ has a magnitude of zero and a negative sign, which means $-0_{10}$. As zero is the only number which cannot have a polarity, $+0$ and $-0$ are somewhat anomalous. However, on increasing the magnitude of the negative numbers we obtain $10000001_2 (-1_{10})$, $10000010_2 (-2_{10})$, etc., up to the largest negative number $11111111_2$ which is equivalent to $-127_{10}$.

Because of the dual representation of zero, this simple signed notation known as one's complement is not suitable for computer arithmetic. A modified signed coding has therefore been developed which has a unique zero representation. This signed number code is known as two's complement. In two's complement positive signed numbers are identical to positive one's complement numbers. At 8-bits resolution they range from $00000000$ $(0_{10})$ to $01111111_2$ $(127_{10})$.

The two's complement code for a negative number is generated by subtracting the magnitude of the negative number from $2^n$, where $n$ is the word length of the number. The two's complement code of $-N$ is defined as

$$-N_{TC} = 2^n - |N| \tag{7.18}$$

The subtraction operation can be simplified because

$$2^n - |N| = \bar{N} + 1 \tag{7.19}$$

where $N$ is the bit-wise inversion of $N$. This can be verified by substituting specific values for $n$ and $N$ in Equations 7.18 and 7.19.

---

**Worked Example 7.2**  Show that the two's complement code for the negative number $-9$ can be calculated by adding $1_2$ to the binary inverse of the number.

*Solution*  Let the two's complement code of $-9_{10}$ be **X**. By definition

See Equation 7.18

$$\mathbf{X} = 2^n - |N|$$

As

$$n = 4 \quad \text{and} \quad |9| = 1001_2$$
$$\mathbf{X} = 2^4 - 1001_2$$
$$= 10000_2 - 1001_2$$
$$= 0111_{TC}$$

Now suppose

$$\mathbf{Y} = \bar{N} + 1$$

As

$$\bar{N} = \overline{1001_2} = 0110_2$$
$$\mathbf{Y} = 0110_2 + 1_2 = 0111_{TC}$$
$$\therefore \mathbf{X} = \mathbf{Y}$$

Hence the two's complement of a negative number can be calculated by adding $1_2$ to its inverse.

---

### Subtraction using two's complement coding

It will now be shown that if the subtrahend is in two's complement code, then subtraction can be achieved with an adder circuit. This has significant implications in computer arithmetic as it means that both addition and subtraction can now be carried out with adder circuits and the subtractor circuits are redundant.

Consider the addition of a two's complement number $\mathbf{B}_{TC}$ to a positive number A as follows:

$$A + \mathbf{B}_{TC} = Z \tag{7.20}$$

But

$$\mathbf{B}_{TC} = 2^n - B$$

Substituting for $\mathbf{B}_{TC}$ in Equation 7.20 gives

$$A + \mathbf{B}_{TC} = A + 2^n - B = A - B + 2^n \tag{7.21}$$

It can be seen therefore that the two complement addition is in fact performing an arithmetical subtraction $A - B$. The addition of $2^n$ to $A - B$ will not affect the

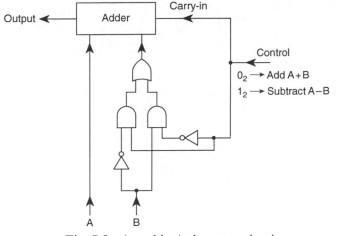

Fig. 7.5   An adder/subtractor circuit.

magnitude of the result and simply appears as an overflow of the most significant cell of the adder and can therefore be ignored.

The two's complement code of the number B is generated by a bit-wise inversion of B and the addition of $1_2$. Hence

$$B_{TC} = 2^n - B = \overline{B} + 1_2$$

Bit-wise inversion is easily achieved by the use of inverter gates and the addition of $1_2$ can be carried out in an adder circuit by setting the carry-in of the least significant stage of the adder to 1. This is of course always set to $0_2$ for normal addition.

Any adder can therefore be upgraded to an adder/subtractor as shown in Fig. 7.5 by including a switchable inverter circuit and a control input.

### BCD addition

BCD coding is often used in calculators because of the ease with which it can be converted into decimal and vice versa in the circuits associated with the input keys and output display. Pure binary adders cannot be used for BCD addition because if the summation of one decimal stage exceeds $9_{10}$, illegal or incorrect BCD codes are generated. If for example the following addition is performed

In BCD code, any decimal digit only depends on four binary digits.

$$9_{10} + 9_{10} = 18_{10}$$

the sum is $18_{10}$ or $10010_2$.

If the numbers are encoded in BCD, the additions in a BCD adder should be

$$1001_{BCD} + 1001_{BCD} = 1,1000_{BCD}$$

$1,1000_{BCD} \equiv 18_{10}$

If, however, the BCD-coded inputs are applied to a binary adder, the following results.

$$1001_{BCD} + 1001_{BCD} = 10010_2$$

$1,0010_{BCD} \equiv 12_{10}$

If the binary output is then interpreted as a BCD code the answer is $12_{10}$ which

159

Fig. 7.6 A binary-coded decimal adder circuit.

is clearly incorrect. Similar errors occur for all inputs producing a sum greater than $9_{10}$ at each decimal place in the addition.

In order to achieve correct BCD addition, a correction facility must be built into the binary adder circuit. A BCD adder is shown in Fig. 7.6. BCD-coded numbers A and B are added together in adder 1 and the sum is monitored by a comparator which produces a signal $F = 1$ if $A + B$ is greater than $9_{10}$. The sum from the first adder, which needs to be 5 bits as its maximum value is $18_{10}$, therefore uses the carry-out as the most significant sum bit.

This 5-bit sum is input to a second binary adder having one input of 5-bits and the other 4-bits resolution. The most and least significant bits of the 4-bit input are permanently set to 0 and bits $b_1$ and $b_2$ are driven from the comparator signal. If the sum $A + B$ does not exceed $9_{10}$, the 4-bit input to adder 2 is therefore $0000_2$ and the output of adder 1 is transferred directly to the output of adder 2. If the comparator does indicate a sum greater than $9_{10}$, the 4-bit input to adder 2 becomes $0110_2$ or $6_{10}$ which is the correction factor to be added to the binary sum $A + B$ in adder 2.

The maximum addition that can take place in adder 2 is $10010_2 + 0110_2 + 1_2 = 11001_2$.

The carry-out is therefore needed in this 5-bit adder circuit. The most significant bit of the output becomes the BCD carry-out and the remainder is the BCD magnitude output, which will always be within the range of one decimal digit, i.e. $0_{10}$ to $9_{10}$.

The 4-bit input adder 2 can only have values of $0_{10}$ or $6_{10}$.

Includes a 1-bit carry-in.

## Multiplication

### Shift-and-add multiplication

In order to understand multiplication in a logic system, the method of arithmetical multiplication must be examined. In the decimal number system, the multiplication of all possible numbers in the range 0–9 must be learnt. These are basic relationships which are defined by the number system. Decimal multiplication is then a process of multiplying in the range 0–9 and shifting the resulting product and adding the result to an accumulator.

In the multiplication

$$A \times B = M$$

the multiplicand is

$$A = a_n \ldots a_2 a_1 a_0$$

the multiplier is

$$B = b_n \ldots b_2 b_1 b_0$$

and the product is

$$M = m_{2n} \ldots m_2 m_1 m_0$$

Note that the resolution of the product must be twice that of the inputs.

Arithmetical multiplication can be carried out in the following steps.

Suppose

$$A = 173_{10} \quad \text{and} \quad B = 263_{10}$$

Step 1   Multiply $A$ by $b_0$      $173_{10} \times 3_{10} = 519_{10}$
Step 2   Shift product $A \times b_0$ zero places to left      $519_{10}$
Step 3   Add shifted product to accumulator $M$      $M = 519_{10}$
Step 4   Multiply $A$ by $b_1$      $173_{10} \times 6_{10} = 1038_{10}$
Step 5   Shift product $A \times b_1$ one place to left      $10380_{10}$
Step 6   Add shifted product to accumulator      $M = 10899_{10}$

The procedure is repeated for each digit in $B$ after which the product $A \times B$ will be in the accumulator $M$.

In the above example, the final value in $M$ will be $45499_{10}$.

This shift-and-add multiplier algorithm can be generalized for any bit $n$ of the multiplier as follows:

Step $n1$   Multiply $A$ by $b_n$
Step $n2$   Shift product $A \times b_n$ $n$ places to left
Step $n3$   Add shifted product to accumulator

*Binary multiplication*

If multiplication of base 2 numbers is required, as will be the case in arithmetic units for computers, the shift-and-add multiplication algorithm can be simplified as there is no actual multiplication operation in the algorithm. If bit $b_i$ of the multiplier is $0_2$ then the multiplication $A \times b_i$ must be zero for all possible values of $A$. If $b_i$ is $1_2$ then the multiplication $A \times b_i$ is equal to $A$. Therefore no extra logical operation or calculation is required in this step of the algorithm. Binary multiplication is therefore only a series of shift-and-add operations and like the subtractor operation, this too can be implemented based on an adder circuit, enhanced with three data stores and a control unit. A block diagram of a binary multiplier is shown in Fig. 7.7.

Fig. 7.7  A binary shift-and-add multiplier.

For example, parallel or look-ahead-carry adder.

The binary adder is the main processor in a multiplier. It can be a ripple-through-carry type or one of the faster adders previously described. The adder resolution needs to be $2n$ bits as the data in register A (the multiplicand) and register B (the multiplier) are $n$ bits. Owing to the increased complexity of the system, a controller unit is necessary to schedule the flow of information. The controller receives the least significant bit $b_0$ from register B via its input K. If $b_0$ is 0, register A is shifted 1 bit left initiated by control signal $C_2$ and register B is shifted 1 bit right through $C_3$. If, however, $b_0$ is 1 then the contents of register A and the accumulator are input to the adder via controls $C_1$ and $C_4$ and the resulting addition stored in the accumulator before the shift controls are applied to registers A and B. The process continues until all bits of the multiplier B have been processed at which point the multiplication of A and B is stored in the accumulator.

---

**Worked Example 7.3**

$11_{10} \times 14_{10}$

Twice the resolution of A and B which are both 4 bits.

$10110_2$ is **A** shifted one place left.

Answer $151_{10}$.

Evaluate the contents of the accumulator after each step in the following multiplication:

$$1011_2 \times 1110_2$$

*Solution*  The multiplicand **A** is $1011_2$ and the multiplier **B** is $1110_2$.
The accumulator must be of 8-bits resolution and is initially $\mathbf{M} = 00000000$.
As $b_0$ is 0, the accumulator contents remain unchanged after the first shift of **B**.
For $b_1 = 1$, $10110_2$ is added to $\mathbf{M}$, giving $\mathbf{M} = 00010110_2$.
For $b_2 = 1$, $101100_2$ is added to $\mathbf{M}$, giving $\mathbf{M} = 01000010_2$.
For $b_3 = 1$, $1011000_2$ is added to $\mathbf{M}$, giving $\mathbf{M} = 10011010_2$.
The accumulator now contains the result of the multiplication.

---

### Counter-based multipliers

The counter-based multiplier is a more simple circuit which does not require any binary adders. It comprises three counters, two of $n$-bits and one of $2n$-bits resolution. The multiplier and multiplicand are loaded into the $n$-bit counters and the $2n$-bit product counter is initially set to zero. The multiplier is then decremented at the same rate as the product counter is incremented. When the

Decrement – count down; increment – count up.

162

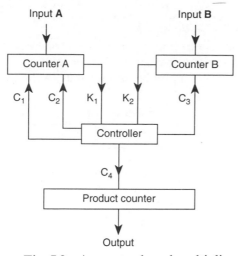

Fig. 7.8   A counter-based multiplier.

multiplier counter reaches zero, the multiplicand counter is decremented by 1, the original value is restored to the multiplier counter and it is again decremented at the same rate at which the product counter continues to be incremented. The process continues until both multiplier and multiplicand counters reach zero, at which point the result of the multiplication is in the product counter.

A block diagram of the counter-based multiplier is shown in Fig. 7.8.

The controller provides the appropriate signals $C_2$, $C_3$ and $C_4$ to increment and decrement the counters. Control $C_1$ restores the multiplier to counter A from an internal register which must be designed into this counter. Signals $k_1$ and $k_2$ enable the controller to detect when the input counters are at zero and hence the end of the multiplication.

Whilst the counter-based multiplier is conceptionally very simple, it has a major drawback in its operating speed. The worst case time to multiply $2n$-bit numbers is $2^{2n}$ counter increments.

Counters are essentially flip-flop circuits with appropriate driving logic and if designed in parallel will have $5\,\mathrm{d}t$ gate delays. This gives a total delay of See page 70 $5\,\mathrm{d}t \times 2^{2n}$. This amounts to $327680\,\mathrm{d}t$ for 8-bit multiplication using this architecture and compares very unfavourably with other arithmetic operations such as $16\,\mathrm{d}t$ for 8-bit ripple-through-carry addition.

**Fast multipliers**

Shift-and-add multiplication is considerably slower than ripple-through-carry addition because in the worst case, $n$ additions are needed in the multiplier, where $n$ is the input resolution. Counter-based multipliers are totally unacceptable for high speed arithmetic as was shown in the previous section. If fast multiplication is needed, then a parallel multiplier or look-up table multiplier must be used. Both these methods give significant improvements in operating speeds over the methods already discussed.

### Parallel multipliers

A multiplier can be built as a canonical logic system, where the product is a direct function of all the inputs and no partial product or shift operation is needed. The truth table for a 2-bit parallel multiplier is shown in Table 7.2.

**Table 7.2   Truth table for a 2-bit parallel multiplier**

| $A_1$ | $A_0$ | $B_1$ | $B_0$ | $M_3$ | $M_2$ | $M_1$ | $M_0$ |
|-------|-------|-------|-------|-------|-------|-------|-------|
| 0 | 0 | 0 | 0 | 0 | 0 | 0 | 0 |
| 0 | 0 | 0 | 1 | 0 | 0 | 0 | 0 |
| 0 | 0 | 1 | 0 | 0 | 0 | 0 | 0 |
| 0 | 0 | 1 | 1 | 0 | 0 | 0 | 0 |
| 0 | 1 | 0 | 0 | 0 | 0 | 0 | 0 |
| 0 | 1 | 0 | 1 | 0 | 0 | 0 | 1 |
| 0 | 1 | 1 | 0 | 0 | 0 | 1 | 0 |
| 0 | 1 | 1 | 1 | 0 | 0 | 1 | 1 |
| 1 | 0 | 0 | 0 | 0 | 0 | 0 | 0 |
| 1 | 0 | 0 | 1 | 0 | 0 | 1 | 0 |
| 1 | 0 | 1 | 0 | 0 | 1 | 0 | 0 |
| 1 | 0 | 1 | 1 | 0 | 1 | 1 | 0 |
| 1 | 1 | 0 | 0 | 0 | 0 | 0 | 0 |
| 1 | 1 | 0 | 1 | 0 | 0 | 1 | 1 |
| 1 | 1 | 1 | 0 | 0 | 1 | 1 | 0 |
| 1 | 1 | 1 | 1 | 1 | 0 | 0 | 1 |

It can be seen that the four functions are

$$M_0 = \Sigma(5,7,13,15)$$
$$M_1 = \Sigma(6,7,9,11,13,14)$$
$$M_2 = \Sigma(10,11,14)$$
$$M_3 = \Sigma(15)$$

These equations can be minimized where appropriate and the 2-bit multiplier built as a two-layer combinational logic system, with an operating time of $2\,\mathrm{d}t$.

Unfortunately parallel multipliers are more expensive to expand than parallel adders. In general an $n$-bit parallel multiplier will contain $2n$ logic functions and each may be dependent on $2n$ variables. The parallel $n$-bit adder only contains $n + 1$ functions. This near exponential increases in complexity with resolution limits parallel multipliers to, at most, 8-bits input resolution.

### Look-up table multipliers

The fastest method of multiplying two numbers is to precompute all possible products and store them in a memory. The address to the memory is formed from the multiplicant and multiplier. If the 2-bit multiplier in Table 7.2 is implemented using a look-up table, the address would be $A_1A_0B_1B_0$ and the 4-bit product $M_3M_2M_1M_0$ is stored in the addressed location. The memory requirements for an $n$-bit look-up table multiplier are $2^{2n} \times 2n$ bits arranged as $2^{2n}$ words of $2n$-bits length.

The address is formed from the two input words to the multiplier.

---

Evaluate the memory requirements for an 8-bit resolution look-up table multiplier.

**Worked Example 7.4**

*Solution*   If the multiplier is 8-bits resolution, the memory address resolution must be 16 bits, i.e. $2n$ where $n = 8$.

This implies a memory of $2^{16}$ locations, i.e. a 64k memory.

Each location must store a product of $2n$-bits resolution, i.e. 16 bits.

The memory therefore needs to be of size

64k $\times$ 16 bits

---

The memory requirement increases exponentially with resolution $n$ and this again limits the size of the multiplier. Some modest savings in memory requirements can be achieved using folding techniques in the memory, which take advantage of the fact that multiplication is commutative. As $\mathbf{A} \times \mathbf{B}$ is identical to $\mathbf{B} \times \mathbf{A}$, only one product needs to be stored for each commutative pair of multiplications. Storage saving of up to 37.5% can be achieved with this method, but this must be offset against the need for a comparator and a more complex address decoder to identify the commutative pairs of multiplications.

This applies to a 2-bit multiplier where only 10 out of the 16 possible input pairs need to be stored.

### Division

A range of division circuits can be designed which use the same principles and ideas found in the family of multiplier circuits.

Division can be achieved using

  (i)  shift-and-subtract
 (ii)  decrementing counters
(iii)  parallel divisors
(iv)  look-up table divisors

As these circuits are all duals of the multiplier circuits which have already been considered in detail, it is left as exercises for the reader to redesign the various multiplier circuits for division.

### A general arithmetic processor

When designing logic systems to perform arithmetic, it is desirable to have a single system which can perform a number of different operations rather than a collection of different circuits, each of which only performs one operation. If a general system is to be designed which performs more than one task, then some form of control information has to be applied to determine which operation the system is to carry out at a given time. The general arithmetic processor, or ALU, is an elementary form of computer and the control strategy in more complex systems becomes the computer program.

Consider the arithmetic unit shown in Fig. 7.9. The main components are an adder circuit and an accumulator A, a register that stores the output of the adder.

There are a number of gates in the unit which are used to control the flow of information through the unit, depending on the value of the control input $C_i$. For example, in gate 1, the output of the accumulator A can only pass through the gate and on to the adder if control input $C_1$ is 1. Other control signals are required by the accumulator (clear and clock) and the adder (carry-in).

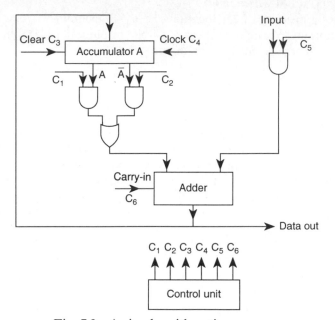

Fig. 7.9   A simple arithmetic processor.

The circuit shown in Fig. 7.9 indicates only one connection line between the various logic components and only one logic gate at each control point. This is something called a bit-slice of the system. It is only a 1-bit processor. In practice the system must be scaled up to the resolution of the data words to be processed, typically 8, 16 or 32 bits. Each communication bus must be of the same resolution as the data word and each bit has its own control gate. The control signals are, however, connected together for each gate at a given control point.

A number of distinct operations that can be carried out in the processor can now be defined. These operations are usually abbreviated as three-letter mnemonics. Eight such operations are shown in Table 7.3.

Each operation in Table 7.3 requires its own distinct set of control signals. For example, STA needs $C_5$ set to 1 to allow the external input to be connected to the adder. If $C_1$ and $C_2$ are 0 then the second addend is zero as no information

can flow from the accumulator to the adder. The output for the adder is therefore equal to the external input and finally a control signal on $C_4$ will load the input into the accumulator store.

**Table 7.3   Eight operations to be performed in an arithmetic unit**

| Operation | Mnemonic |
|---|---|
| Input | INP |
| Store input | STA |
| Add input to accumulator | AIA |
| Invert accumulator | IVA |
| Subtract accumulator from input | SAI |
| Output accumulator | OPA |
| Output inverse of accumulator | OIA |
| Increment accumulator | ICA |

The SAI operation – subtract accumulator from input – is the most complicated of the eight instructions. It involves generating the two's complement code of the number in the accumulator. This is achieved by inverting the accumulator (control $C_2$). The addition of $1_2$, which is required to complete the two's complement code, is carried out by setting the carry-in to the adder control $C_6$ to $1_2$. The two's complement code of the number in the accumulator is then added to the minuend applied to the data-in terminal and this performs the desired subtraction.

The full set of control signals for each of the unit's eight operations is given in Table 7.4.

**Table 7.4   Control signals for the eight ALU operations**

| Operation | Controls | | | | | |
|---|---|---|---|---|---|---|
|  | $C_1$ | $C_2$ | $C_3$ | $C_4$ | $C_5$ | $C_6$ |
| Input | 0 | 0 | 1 | 0 | 1 | 0 |
| Store input | 0 | 0 | 1 | 1 | 1 | 0 |
| Add input to accumulator | 1 | 0 | 0 | 1 | 1 | 0 |
| Invert accumulator | 0 | 1 | 0 | 1 | 0 | 0 |
| Subtract accumulator from input | 0 | 1 | 0 | 1 | 1 | 1 |
| Output accumulator | 1 | 0 | 0 | 0 | 0 | 0 |
| Output inverse of accumulator | 0 | 1 | 0 | 0 | 0 | 0 |
| Increment accumulator | 1 | 0 | 0 | 1 | 0 | 1 |

### Design of the control unit

The simple arithmetic unit described in the previous section has six control lines. This means that there are 64 possible operations that can, in theory, be performed. Eight operations have been identified as being useful for arithmetic processing. There may be others of some value but the majority of combinations of control signals do not lead to any meaningful operation of the unit.

This is all possible combinations of control signals, i.e. $2^6$.

**Table 7.5  A 3-bit coding of the instruction set**

| Mnemonic | A | B | C |
|----------|---|---|---|
| INP | 0 | 0 | 0 |
| STA | 0 | 0 | 1 |
| AIA | 0 | 1 | 0 |
| IVA | 0 | 1 | 1 |
| SAI | 1 | 0 | 0 |
| OPA | 1 | 0 | 1 |
| OIA | 1 | 1 | 0 |
| ICA | 1 | 1 | 1 |

$Log_2\, 8 = 3.$

If the arithmetic unit is limited to the eight operations in Table 7.4, then each operation could be defined by a 3-bit code applied to the input of the control unit rather than a 6-bit code applied directly to the arithmetic unit. It is obviously easier to program the device with a sequence of 3-bit codes rather than 6-bit codes. In order to do this the eight operations have to be allocated their own 3-bit codes. Suppose the allocation of a 3-bit code ABC is as shown in Table 7.5

A decoder table can now be generated, which is implemented in the control unit. The simple ALU used in this chapter produces six control signals $C_1$ to $C_6$ and has three inputs ABC, as shown in Fig. 7.10.

The control unit therefore contains six combinational logic functions each of three inputs and the truth table for the complete unit is given in Table 7.6.

The individual functions can now be specified. For example,

$$C_1 = f(ABC) = \Sigma(2,5,7)$$

Using a K-map to minimize the function gives

Dotted loops apply to 2nd canonical form

Fig. 7.10  The control unit.

**Table 7.6 Truth table for the control unit**

| Input | | | Control signals | | | | | |
|---|---|---|---|---|---|---|---|---|
| A | B | C | $C_1$ | $C_2$ | $C_3$ | $C_4$ | $C_5$ | $C_6$ |
| 0 | 0 | 0 | 0 | 0 | 1 | 0 | 1 | 0 |
| 0 | 0 | 1 | 0 | 0 | 1 | 1 | 1 | 0 |
| 0 | 1 | 0 | 1 | 0 | 0 | 1 | 1 | 0 |
| 0 | 1 | 1 | 0 | 1 | 0 | 1 | 0 | 0 |
| 1 | 0 | 0 | 0 | 1 | 0 | 1 | 1 | 1 |
| 1 | 0 | 1 | 1 | 0 | 0 | 0 | 0 | 0 |
| 1 | 1 | 0 | 0 | 1 | 0 | 0 | 0 | 0 |
| 1 | 1 | 1 | 1 | 0 | 0 | 1 | 0 | 1 |

Hence

$$C_1 = AC + \overline{A}\,B\overline{C}$$

in the 1st canonical form or

$$C_1 = (A + \overline{C}).(\overline{A} + C).(B + C)$$

in the 2nd canonical form.

Design the remaining functions $C_2$ to $C_6$ for the control unit.

**Exercise 7.3**

## Programming the system

Since an arithmetic unit has been specified and a set of instructions identified, the system can now be programmed to perform arithmetic.

Consider the calculation

$$S = (A - B) \mid C$$

This can be carried out by first loading B into the accumulator, subtracting the accumulator from input A and then adding input C to the accumulator and finally outputting the accumulator.

The sequence of instructions for this calculation is

| Instruction | Comment |
|---|---|
| INA | Apply B to input |
| STA | |
| INA | Apply A to input |
| SAI | |
| INA | Apply C to input |
| AIA | |
| OPA | |

The program has to be stored in a memory. In a computer this could be the program store. Each instruction is accessed sequentially, and controlled by a program counter, which is also used to synchronize the input of data. By changing the sequence of instructions, other calculations can be carried out.

The arithmetic unit developed in this chapter is very simple and limited, but it does serve to illustrate the principles of a programmable system. Practical ALUs contain many more registers and processing units and sometimes have over 100 control lines. The instruction set which defines the operation of the unit could amount to 30 to 50 different entries. The reader should be aware that three very important subjects in computer science have been demonstrated here in an elementary form. The compilation of a list of mnemonic instructions to configure the unit is an elementary form of assembler language programming; the application of the appropriate binary signals within the ALU is similar to machine code programming. If the logic functions in the control are changed in order to optimize or customize the device, this is a form of microprogramming.

## Summary

Logic systems can be designed to perform arithmetic operations. The simpler systems, in terms of the amount of logic required, tend to be slow in operation, but can easily be expanded to process higher resolution data. Parallel arithmetic units are much faster but require more complex functions. The complexity is an exponential function of the resolution and this limits the resolution of parallel arithmetic units in most cases to a maximum of 8 bits.

The two's complement method of coding negative numbers allows subtraction to be carried out in an adder circuit. Multiplication can be implemented using a shift-and-add algorithm and division uses shift-and-subtract operations. The principal arithmetic operations can therefore be carried out in an arithmetic unit, based on an adder circuit. The sequencing of the control signals to the ALU leads to the concept of a program and the design of the control unit enables the system to be reconfigured or optimized for specific processing applications.

## Problems

7.1 Design a full adder cell using only the minimum number of NAND gates.

7.2 Design a full adder cell using only the minimum number of NOR gates.

7.3 Evaluate the carry function $C_2(out)$ for cell 2 of a look-ahead-carry-adder. (Note that the least significant cell is cell 0.)

7.4 Convert the following numbers to two's complement

    (i) $-41_{10}$

    (ii) $-101101_2$

    (iii) $-57_8$

    (iv) $-E4_{16}$

7.5 Design, using canonical form logic, a 2-bit resolution parallel subtractor with two's complement output coding.

7.6 Redesign the subtraction of Problem 7.5 using memories to implement the logic functions.

7.7 Design a BCD adder circuit using a memory look-up table. What size of memory is required if the inputs are applied directly to the memory address lines? How could the memory requirement be reduced?

7.8 Design a shift-and-add multiplier in block schematic form and specify a control sequence to carry out the multiplication of two numbers.

7.9 Design a combinational logic processor to generate the square of a 4-bit input.

7.10 Show that subtraction can be performed by two's complement addition when the minuend is negative. How can negative overflow be detected in this processor?

# 8 Practical digital circuits

In this final chapter, we shall concentrate on some of the practical aspects of a digital logic system. The different logic families will be identified and their characteristics examined. Practical constraints that must be observed when a logic system is put together will be discussed. Furthermore, a working digital system does not exist on its own. It has to be interfaced with other systems, such as industrial processes, machines and people. Interfacing the digital computer with the analogue world is therefore an important area, and the design of data converters for this purpose will be outlined.

## Logic families

Integrated circuit logic families can be divided into two broad groups: the bipolar and the metal oxide semiconductor (MOS) families. Both rely on the switching of transistors between two discrete states in order to represent the logical behaviour of a function. The bipolar logic families are based on the bipolar transistor, whereas the MOS systems use the unipolar field effect transistor as their component. Both types are suited to integrated circuit manufacture.

The two states are usually cut-off and saturation.

## Bipolar transistor logic

The earliest electronic logic was based on diodes that could represent the logical constants by being either forward or reverse biased. Diode logic circuits cannot, however, perform the NOT operation, and there is a loss of voltage at each gate, owing to the contact potential of the diode. Therefore, if a circuit has several gates following one another, it is possible that the voltages representing the logical constants 0 and 1 could overlap and the system fail. The transistor can also act as a two-state switch, and as an amplifier with 180° phase change. All these properties are needed in a logic circuit. The two distinct states represent the logic constants: the amplifying property enables the voltage levels representing the states to be maintained, and the 180° phase shift acts as an inverting function.

Contact potential approx. 0.3 V for germanium and 0.7 V for silicon diodes.

### Resistor transistor logic

The earliest form of transistor logic was resistor transistor logic (RTL) where input variables were applied via biasing resistors to the base of a transistor. The values of the resistors were chosen so that the inputs would, depending on their voltage levels, either saturate or cut off the transistor.

Beware, RTL also stands for 'register transfer language'.

### Diode transistor logic

Diode transistor logic (DTL), where the logical AND or OR operations were performed by the diodes and the transistor acted as a voltage amplifier and inverter, appeared in the early 1960s. Compared with RTL, DTL offered faster switching speed, as the diode input circuits provided lower driving resistance to transistors.

### Transistor transistor logic

In 1964, Texas Instruments introduced their 7400 series of transistor transistor logic (TTL), which has dominated the logic component market for the past two decades and is still widely used today. The principal features of a TTL gate is a multiple emitter input transistor that is functionally equivalent to the input diodes in a DTL circuit.

Since its inception, the TTL logic structure has been continuously developed to meet more stringent speed and power consumption standards. The following TTL series are currently available:

(i) High speed TTL where the circuitry is basically the same as the standard TTL but the resistors within the circuits have been reduced in value, resulting in faster switching at the expense of increased power consumption. — 74H00 Series

(ii) Low power TTL where the resistor values have been increased giving reduced power consumption at the expense of longer propagation times. — 74L00 Series

(iii) Schottky TTL which has a Schottky barrier diode connected between the base and collector of every transistor in the gates. A Schottky barrier diode has a contact potential of about 0.25 V, compared with 0.7 V for a silicon diode, and prevents the transistor saturating heavily. It can, therefore, change state much faster and is the highest speed TTL available. — 74S00 Series

(iv) Low power Schottky TTL, where the use of Schottky diodes and high resistor values lead to low power consumption, but the reduction in speed is less than in the low power range, owing to the diode clamps. — 74LS00 Series

### Emitter coupled logic

In the standard TTL ranges, the transistors are saturated during part of their operating cycle. If the saturation is restricted, as in the case of Schottky TTL, the switching speed is increased, resulting in faster devices. An alternative non-saturating bipolar family known as emitter coupled logic (ECL) is available, where the switching is controlled by current generators. When the transistor is switched from the cut-off to conducting state, the resulting collector current is controlled by the fixed current injected into the base which is less than the saturation current $I_c(\text{sat})$. Consequently, ECL provides very high speed switching,

Sometimes known as 'current mode logic' (CML).

173

**Table 8.1  Typical power and propagation data for bipolar logic families**

| Name | Series number | Propagation (ns) | Power (mW) |
|------|---------------|------------------|------------|
| Standard TTL | 7400 | 10 | 10 |
| High speed TTL | 74H00 | 5 | 25 |
| Low power TTL | 74L00 | 35 | 1 |
| Schottky TTL | 74S00 | 3 | 25 |
| Low power Schottky TTL | 74LS00 | 10 | 2 |
| Emitter coupled logic | ECL | 2 | 25 |
| $I^2L$ lowest power | | 100 | $5 \times 10^{-6}$ |
| $I^2L$ fastest speed | | 5 | 5 |

being marginally faster than Schottky TTL and having approximately the same power consumption.

ECL gates have different voltage levels for the logic states ($-0.8$ V and $-1.7$ V) compared with TTL (0 V to 0.4 V and 2.4 V to 5.0 V) and are not therefore directly compatible. An ECL gate, however, usually generates both normal and inverted outputs and thereby eliminates the need for inverters.

*Integrated injection logic*

Integrated injection logic ($I^2L$) is the latest bipolar family. It offers a range of speed/power consumption trade-offs and can operate in both saturated and unsaturated modes. High packing density is achieved because biasing resistors are external components not included within the chips. It is likely, therefore, that $I^2L$ will have a significant effect on VLSI design.

The power/speed characteristics of bipolar transistor logic families are summarized in Table 8.1.

**Unipolar logic families**

Unipolar logic families are based on the field effect transistor which requires a metal electrode separated from a semiconductor channel by an oxide insulating layer. This is the MOS fabrication technology, and the individual transistors are often referred to as MOSFETs.

MOS logic devices, unlike bipolar logic (except for $I^2L$), do not require internal resistors on the chips, and can hence be manufactured at a high packing density. The fabrication process is relatively simple, and power consumption is low. MOS ICs, however, have relatively slow operating times and are therefore not suited to logic systems where very fast switching is essential.

*P-MOS logic*

P-MOS logic is a family of MOS integrated circuits. The semiconductor material is p-doped and the majority carriers are therefore holes. P-MOS is suited to large scale integration and has a greater packing density capability than bipolar transistor logic.

A resistor within an integrated circuit takes up 10 times more chip area than a transistor.

MOSFETs: Metal Oxide Silicon Field Effect Transistors.

Ideal for battery-operated systems.

## N-MOS logic

N-MOS Logic is basically the same structure as P-MOS, except that it uses n-doped semiconductor material. The circuit carriers in N-MOS are free electrons, which are more mobile than larger positive charges, and this results in faster switching times compared with P-MOS devices.

## CMOS logic

Complementary metal oxide semiconductor logic uses both p- and n-type channels in the same circuit. It is faster than P-MOS and N-MOS and requires considerably less power than the low power TTL series. CMOS is, however, still inferior to standard TTL in terms of operating speed. MOS circuits can operate off a wide range of supply voltages. If CMOS is being used alongside TTL, a 5 V supply would be used. However, when used alone, it will operate satisfactorily over the range 3 to 20 V. The input levels are dependent on the supply voltage $V_s$ and are approximately 30% and 70% of $V_s$ for the low and high states, respectively. Power dissipation is also dependent on the supply voltage, but is still extremely low compared with other families.

## Silicon on sapphire

SOS is a recent development in MOS technology where the oxide insulating material is replaced by sapphire. This has the effect of reducing internal capacitance and thereby increases the speed of the logic.

For electronic circuit analysis of bipolar and unipolar transistor logic gates refer to Chapters 6 and 7 of Richie, G.J. *Transistor Circuit Techniques – discrete and integrated* (3rd edn, Chapman & Hall, 1993).

## Practical constraints on logic circuits

All logic families are subject to certain constraints in use. The main ones are identified below, but the designer must always familiarize him- or herself with the manufacturer's data sheet when using particular products.

## Fan-out

Fan-out is the number of logic inputs that can be driven from the output of a logic gate. A typical fan-out for a TTL gate is 10. However, with CMOS logic, owing to the very high input impedance of the field effect transistor ($10^{11}\ \Omega$), the fan-out is potentially infinite. In practice, however, owing to capacitive loading of the driving output by the inputs, the fan-out is generally restricted to a maximum of 50.

Special drivers are available that will drive more than 10 gates.

## Fan-in

Fan-in is the number of logic outputs that can be connected to a logic gate input. A fan-in greater than 1 was possible with the now obsolete RTL, where every output was fed via its own external resistor to the input in question. Standard TTL outputs, must NEVER be connected together, as the internal transistors will be overloaded if one output is low and another is high. The fan-in of a standard TTL gate is therefore 1. Special open collector gates can be obtained, where the output transistor load is not included on the chip. It has to be supplied externally.

VERY IMPORTANT.

The outputs of open collector gates can be connected together and a 'wired-AND' operation obtained. The output across the external load resistor is high only if all the gate outputs are high. If one or more gate inputs are low then the 'wired-AND' output goes low.

*Unused inputs*

For example, to obtain a two-input NAND function from a three-input gate, the third input must be held at logical 1.

If one or more inputs to a gate are not needed for a particular logic function, the unused inputs should be connected to either logical 1 or logical 0, depending on the function being performed. Alternatively, they may be connected in common with a used input line to form a single input. Unused inputs must NOT be left unconnected, as they are highly susceptible to noise and pick-up which can give rise to incorrect logic behaviour.

An unconnected TTL input, in the absence of noise or pick-up, will appear to be high (logical 1 with positive logic coding) but in a practical circuit this cannot be guaranteed. In CMOS circuits a floating input could bias both p and n channels towards their conducting states and lead to excessive power consumption.

*Handling precautions*

A rather surprising problem arises with CMOS devices, which can result in the circuits being damaged unless suitable precautions are taken. Owing to the extremely high input impedance that can be tens of thousands of megaohms, static charge can build up on the input terminals of a CMOS chip, which can lead to breakdown within the circuit. Great care should be taken when handling CMOS components, and the devices should always be stored with their terminals embedded in conducting foam, to prevent this static build-up.

**Interfacing CMOS and TTL**

In many applications the designer may wish to use CMOS logic, where low power requirements are a priority, and TTL for the parts of the system working at high speed. A system may therefore contain both CMOS and TTL families. The supply rail must be $+5$ V for TTL, whereas CMOS requires a supply voltage of between 3 V and 20 V. A common $+5$ V rail will therefore satisfy both families.

*CMOS output driving TTL input*

A high output from a CMOS gate can drive a TTL input, but when the CMOS is low, the input current that flows back from the TTL gate through the CMOS output to earth (see Fig. 8.1) is sufficiently large to cause a potential drop across the CMOS output resistance that will appear as a logical high to the TTL input. A typical value for the output resistance of a CMOS gate is 2 k$\Omega$ and the current drain of a TTL input 1.5 mA, giving a voltage of the order of 3 V at the TTL input, which would appear to be a logical high. Standard CMOS cannot therefore drive standard TTL. A special CMOS buffer must be used that is designed to sink the current flowing back from the TTL input. CMOS can, however, drive the low power TTL families as the input currents are much smaller compared with the standard series.

Low power (74L00) and low power Schottky (74LS00).

Fig. 8.1 (a) CMOS driving TTL. (b) Equivalent circuit of path A → B → C in (a). Point B cannot fall below 3.0 V.

*TTL output driving CMOS inputs*

TTL will drive CMOS because its low output is approximately 0.4 V on open circuit and this is within the range 0–30% of $V_s$ (0–1.5 V) acceptable as a low input for CMOS. The open circuit high output for TTL is about 3.6 V which is just within the range of $> 70\%$ of $V_s$ (3.5–5.0 V) acceptable as a high input to CMOS. The TTL output is, however, dangerously close to the lower limit of the range and a mere 0.1 V noise could cause a malfunction. In practice, therefore, external pull-up resistors are used to increase the high output voltage of the TTL gate and hence give greater reliability (Fig. 8.2).

CMOS input impedance is extremely high and therefore appears as an open circuit to the TTL.

$V_s = +5$V

Typical resistor value is 1kΩ

Fig. 8.2 A pull-up resistor used when TTL drives CMOS.

**Practical memory circuits**

Memory devices have two very important roles in logic design. They are storage devices and can also act as programmable universal logic functions. Consequently memory technology has received much attention during recent years. Memories can be divided into two broad groups – the random access memory or RAM, and the read only memory or ROM.

The RAM is a volatile memory. Its contents are lost when the power supply is turned off. The time taken to read the contents of a storage location is of the same order as the time needed to write information into the store. Two types of RAM are available – static and dynamic. In the static RAM, storage elements are flip-flops and the stored data remains constant, provided power is supplied to the chip. The dynamic RAM relies on stored charge on capacitors within the integrated circuit. Stored charge represents logical 1, whereas no charge indicates logical 0. Unfortunately, capacitors will not hold charge indefinitely. It will gradually leak away. Consequently a dynamic memory must be read periodically and those capacitors storing logical 1 must be recharged. This process is called 'refreshing' and must be carried out every few milliseconds Nowadays, dynamic RAMs have refresh logic included on the memory chips and no additional external circuitry is required. Dynamic RAMs require less circuitry per bit than static RAMs. Approximately four times more memory can be fabricated on a given area of silicon chip, if dynamic rather than static storage is used.

Compared with a read or write time of 100 nanoseconds.

The ROM is, by contrast, a non-volatile memory. The stored data is NOT lost when the power supply is turned off. The write procedure is very much longer than the read operation. Write time to read time ratios may be as high as $10^9$ for the ROMs that can be programmed by the user.

ROM: Read Only Memory.

It may take several minutes to write into a ROM.

The simple ROM consists of an XY matrix of wires. The X wires are addressed by a decoder and the Y wires provide output data as described in Chapter 6. The data is stored by making connections between the XY intersection during manufacture.

To store a 1, connect X to Y. To store a 0, leave XY open circuit.

A diode or a transistor.

Programmable ROMs or PROMs have a coupling device in series with a fusible link at each XY intersection. To program the device, the user must pass a large current through each link where a logical 0 needs to be stored. The link is thereby burned or open-circuited, causing the output to go low.

EPROMs are erasable PROMs. The conduction paths and hence the stored data in these memories are defined by the build-up of charge within the gates. This charge can be released and the memory erased if the device is exposed to high intensity ultraviolet radiation for several minutes. The device can then be reprogrammed.

The most recent addition to the memory range is the EAROM or electrically alterable ROM. It is user programmable and electrically erasable, and thereby dispenses with the ultraviolet radiation source required by the EPROM.

**Semi-custom logic**

Current trends in logic design are towards fabricating logic systems within integrated circuits rather than having systems of discrete components interconnected on a printed circuit board. Multiplexers provide a means whereby any combinational logic function of $n$ variables can be set up by the user without resorting to discrete gates. These devices are often marketed as universal logic modules or ULMs.

Memories can be used as programmable logic functions. They are capable of holding much larger functions than the multiplexers, but the initial set-up procedure is slightly more complicated, requiring each individual minterm to be applied to the address field, and the desired output stored in the accessed location. Logic systems of more than one function can be programmed into standard PLAs and PALs provided they have the capacity to hold the logic system by having enough inputs, product lines and outputs.

There is currently, however, a substantial growth in semi-custom logic, where the designer produces a unique silicon chip for the system to be marketed. The uncommitted logic array, or ULA, now being produced by several manufacturers comprises a collection of gates prefabricated onto a chip. The final mask which defines the connections is not included in the initial fabrication process. It is specified by the user and applied to the gate array by the manufacturer prior to delivery. Gate arrays can contain up to 50000 gates and the turn-round time between specifying the connections to the manufacturer and taking delivery of the customized chip is usually a few weeks. As a means of producing a special purpose chip ULAs offer ease of design and relatively low development cost.

The basic ideas behind gate arrays can be illustrated in the following worked example.

---

**Worked Example 8.1**

Design a decoder which will convert a 4-bit binary word (ABCD) into one of 16 output lines ($D_0$–$D_{15}$).

*Solution*   Each output of a decoder is generated by a four-input AND gate where $D_0 = \overline{A}\,\overline{B}\,\overline{C}\,\overline{D}$, $D_1 = \overline{A}\,\overline{B}\,\overline{C}D$, through to $D_{15} = ABCD$.

Now taking $D_0$ and $D_1$ and considering part of an uncommitted gate array, four inputs are required and two output lines. Suppose the array has a number of inverters and two input AND gates, initially with their inputs and outputs unconnected, but all the power connections are present. Pads are available on the array to make the external input and output connections. The array prior to customizing is shown in Fig. 8.3.

The gate inputs and outputs in the array have reference labels, as do the external inputs and output. In order to build up the function $D_0$ the following connections are required:

A–E, B–F, C–G, D–H

These provide inversions of the inputs:

S–I, T–J, U–K, V–L

Fig. 8.3   Part of a gate array prior to customizing.

179

These provide the sub-functions $\overline{A}\,\overline{B}$ and $\overline{C}\,\overline{D}$;

W–M and X–N

These provide the function $D_0 = \overline{A}\,\overline{B}\,\overline{C}\,\overline{D}$. Finally

Y–$D_0$

This provides the output connection.

It is left as an exercise for the reader to show that the following connection list enables the $D_1$ to be generated:

U–O, D–P, Z–Q, W–R and AA–$D_1$

---

In this example the number of connections is quite large for what is a very simple system. In larger practical designs it is impossible to generate the routing list without some computer assistance, and computer-assisted design packages are available for automatic routing of the connections, timing and fan-out checks, and test patterns generation for fault diagnosis. The fabrication technology for gate arrays is generally CMOS.

The most complex form of programmable logic which is available to the designer is the standard cell array. In this form the designer has a library of standard cells or functions on which to base chip design. A complete set of masks based on these cells is then produced and a fabrication run can then be set up for individual customers. Cell arrays offer an even greater degree of customizing than gate arrays, as the designer only includes those cells required by the system. The development cost is higher than for gate arrays, but because standard cells with readily available masks are used, the cost is much less than when a customized chip is designed from first principles. There are also significant savings in development time if cell libraries are available. CAD facilities for all array designs do need to be comprehensive. In particular logic simulation is essential in order to prove the design in software before committing it to silicon.

**Interfacing the digital system to the real world**

We saw in Chapter 1 that the real world invariably appears to be continuous or analogue, whereas operations in a digital system are, by definition, discrete. In any practical application of a digital system, an interface with continuous or analogue processes is needed. This interface is achieved by using analogue to digital converters (ADCs) at the input and digital to analogue converters (DACs) at the output of a digital system.

*Digital to analogue converter*

A simple DAC can be built up from a series of weighted resistors. Taking a 4-bit binary word

$\mathbf{B} = b_3 b_2 b_1 b_0$

the most significant bit is connected to a resistor of value $2R$, bit $b_2$ to a $4R$ resistor, $b_1$ to an $8R$ resistor and so on. The value of the resistor doubles for each less significant bit. If any bit is at logical 1, a voltage is applied across its resistor,

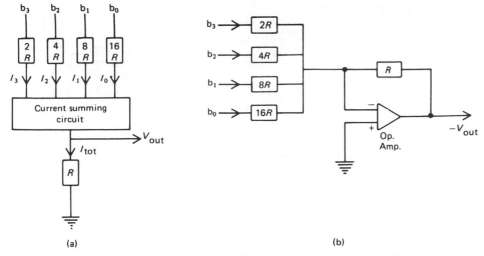

Fig. 8.4   A weighted resistor DAC. (a) Block diagram. (b) Circuit diagram using an operational amplifier.

and a current inversely proportional to the value of the resistor flows. An operational amplifier circuit can be used to sum the currents and produce a voltage proportional to the total current flowing. The magnitude of the voltage will therefore be directly proportional to the value of the binary number **B**. The block diagram of a weighted resistor DAC and its circuit diagram, using an operational amplifier, are shown in Fig. 8.4. The weighted resistor DAC, although very simple, has two major drawbacks: a wide range of resistor values is needed, and the ratio of the smallest to largest resistor is an exponential function of the number of bits:

$$\frac{R_0}{R_n} = 2^n \tag{8.1}$$

The resolution is $n + 1$ bits

So, if a 16-bit converter was constructed and the resistor for bit $b_{15}$ was 2 kΩ, the least significant bit resistor would need to be $2^{15} \times 2$ kΩ or 65.536 MΩ. Furthermore, the resistors must be sufficiently accurate so that the contribution to the total current from the least significant bit is not masked by the current arising from inaccuracies in the most significant bit resistor. For a 16-bit converter the accuracy of the resistor $R_{15}$ must be better than 0.0015% and, in practice, it is impossible to manufacture resistors to such a high accuracy over a wide range of values. Hence weighted resistor DACs have limited practical use, except for very low resolution conversions.

Design a 4-bit weighted resistor DAC and calculate the maximum tolerance on the resistors if the converter is to be accurate to the nearest least significant bit.

**Exercise 8.1**

An alternative approach to digital to analogue conversion is found in an $R/2R$ resistor ladder network as shown in Fig. 8.5. This converter only requires two different resistor values and the network can be fabricated as an integrated circuit. Every bit in the binary word **B** controls a switch. The most significant bit is applied to $SW_3$ through to the least significant bit controlling $SW_0$.

Fig. 8.5   An $R/2R$ resistor ladder network digital to analogue converter.

The input of an inverting
operational amplifier is a virtual
earth. See Horrocks, D. H.
*Feedback Circuits and Op. Amps*
(2nd edn, Chapman & Hall, 1990).

The resistor network is a current dividing circuit and each switch directs the current in its branch either to earth if the input driving the switch is at logical 0 or to the current summing operational amplifier if at logical 1. The operational amplifier input (node F) appears to be an earth as far as the resistor network is concerned. Hence the network can be analysed by assuming that the $2R$ resistors are connected to earth, regardless of the switch settings.

At each node (A,B,C,D) in the ladder network, the resistance is $2R$ when looking either towards the switches or towards earth point E. Hence any current flowing into the node will be divided equally. Only half will flow towards the switch and the other half flows towards the next node, where it is again divided. If the current flowing from the power supply is $I$, then $I/2$ flows to $SW_3$, $I/4$ to $SW_2$, $I/8$ to $SW_1$, and so on. The digital word **B**, which controls the setting of the switches, will determine whether these current contributions are summed or returned to earth. Hence the total current entering the summer is determined by the digital input and appears as a proportional analogue voltage at the output. The resolution of the converter can be increased by extending the $R/2R$ ladder network.

### Analogue to digital converter

One of the most widely used analogue to digital conversion techniques is the successive approximation method, where a voltage is generated by a digital to analogue converter and continuously compared with the analogue input. When the voltages are equal, the digital equivalent of the analogue input is the value of the final input to the DAC.

An analogue input is always a
voltage.

Fig. 8.6   A successive approximation analogue to digital converter.

A block diagram of the successive approximation ADC is given in Fig. 8.6. The controller can be a simple cyclic binary counter and a digital output occurs every time $V_D = V_A$. This method is relatively slow, requiring $2^n$ internal digital to analogue conversions for each analogue to digital conversion. The speed can be increased by incorporating an up/down counter in the controller and setting the direction of the count according to the relative values of $V_D$ and $V_A$ as measured by the comparator. If the analogue input is increasing, the counter is incremented, and if decreasing, it must be decremented from its previous conversion value. In general, the number of steps between conversions will be much less than with the cyclic counter, leading to potentially faster conversion times. The number of steps between conversions will, however, vary with the rate of change of the analogue input. If the worst case conditions are allowed, where the analogue input swings from zero to its maximum value between consecutive conversions, then the up/down converter version will not be any faster than the cyclic controller.

An order of magnitude increase in speed can be obtained by replacing the counter with a processor which initially sets the most significant bit of the DAC to logical 1 and compares $V_D$ with $V_A$.

Initially all bits are at 0.

If $V_D > V_A$ then the most significant bit of the DAC is reset to 0.
If $V_D < V_A$ then the most significant bit remains at 1.

The next bit is set to 1 and its final value determined by the $V_D/V_A$ comparison. The procedure is repeated for each bit driving the DAC. The digital equivalent of the analogue input is the output of the controller after the least significant bit has been processed. This method requires $n$ internal comparisons for each analogue to digital conversion, where $n$ is the resolution in bits. The operating time in this method is independent of the value or rate of change of the analogue input.

Compare with $2^n$ operations for the counter-based converter.

Another commonly used technique is found in the dual-slope integrating ADC. The essential components are shown in Fig. 8.7. The analogue input when applied to an integrator starts an $n$-bit binary counter, driven from an accurate clock. The input voltage is continuously integrated until the $n$-bit counter overflows and returns to zero after $2^n$ clock pulses.

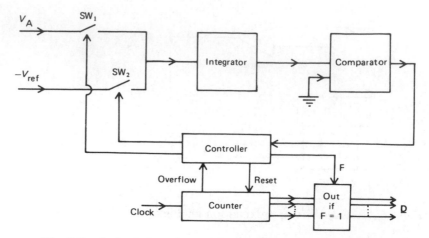

Fig. 8.7   A dual-slope integrating analogue to digital converter.

If the clock period is $\Delta t$ and the analogue input is $V_A$, the output of the integrator when the counter returns to zero will be

$V_A$ is constant during conversion.

$$V_I = \int_0^{2^n \Delta t} V_A \, dt = V_A 2^n \Delta t \tag{8.2}$$

$V_A$ is then switched out of the integrator, and a negative reference voltage $(-V_{ref})$ switched in, causing the integrator output to decrease. When the integrator output is zero (as detected by the comparator) the counter is halted. If $D$ is its final value, the integrator output can be expressed as

$$V_A 2^n \Delta t - V_{ref} D \Delta t = 0 \tag{8.3}$$

Hence

$$V_A = V_{ref} D / 2^n \tag{8.4}$$

$V_{ref}/2^n$ is constant.

So the final value of the counter is directly proportional to the digital equivalent of the analogue input $V_A$.

All the ADCs we have examined so far are relatively slow compared with DACs. The digital output is continuously built up and compared with the analogue input, and this is essentially a serial process. A parallel method is now feasible owing to the development of low cost microcircuits whereby the input voltage is compared with reference voltages for each and every possible digital output. These devices are called 'flash converters' and their operation can be explained as follows.

Suppose we have a very simple converter that has 2-bits resolution and the analogue input voltage is in the range $0 \leqslant V_A \leqslant 4.0\,V$. The digital outputs 00, 01, 10 and 11 indicative whether $V_A$ is closest to 0 V, 1 V, 2 V or 3 V, respectively. By choosing reference voltages at 0.5 V, 1.5 V and 2.5 V and comparing the input $V_A$ with each of these voltages in parallel, the digital equivalent of the input can be obtained. A comparator output at 0 implies $V_A$ is less than $V_{ref}$, and at 1 indicates $V_A$ is greater than $V_{ref}$. The reference voltages can be generated from a precise resistor network as shown in Fig. 8.8 and the comparator outputs must be decoded into pure binary as shown in Table 8.2.

Fig. 8.8   A parallel or flash analogue to digital converter.

Flash converters are extremely fast as the comparisons are all performed at the same time. The reference input can be adjusted externally according to the maximum value of $V_A$ to be encountered. An $n$-bit converter requires $2^n - 1$ comparators so the hardware requirements for high resolution converters can become excessive. This method is therefore only used where high speed at low resolution is required. A typical application is the digitization of television pictures where the conversion frequency is of the order of 10 MHz.

Typically 4-bits resolution.

**Table 8.2   A flash converter output decoder table.** Comparator at 0 implies $V_A < V_{ref}$ and at 1 implies $V_A > V_{ref}$

| $V_A$ | Comparators | | | Decoder output **D** |
|---|---|---|---|---|
| | 0.5 V | 1.5 V | 2.5 V | |
| $0\,V \rightarrow 0.49\,V$ | 0 | 0 | 0 | 0 0 |
| $0.5\,V \rightarrow 1.49\,V$ | 1 | 0 | 0 | 0 1 |
| $1.5\,V \rightarrow 2.49\,V$ | 1 | 1 | 0 | 1 0 |
| $2.5\,V \rightarrow 3.99\,V$ | 1 | 1 | 1 | 1 1 |

**Problems arising from logic usage**

The reader should now be familiar with the fundamental design principles for a digital system. The end product of any successful design exercise is a working system, and this goal may be unattainable if certain practical precautions are not observed. In this final section, we will consider problems that arise when a logic system is actually built up on a printed circuit board.

## Power supply decoupling

The wires and leads interconnecting logic components in a system will have a small but finite inductance. When a logic gate switches, a narrow spike of current flows in the power supply lines. A changing current in an inductor will generate a voltage

$$V = \frac{L \, di}{dt} \tag{8.5}$$

where $L$ is the inductance of the wire. Suppose a current of 20 mA flows for 100 nanoseconds when a gate switches. If the distributed inductance is $10^{-4}$ henrys, then a voltage of 20 V is induced on the power rails. This voltage spike can cause serious malfunction and possible damage in the logic circuits. The induced voltage can be removed by connecting small radio frequency capacitors between the supply input and earth pins of each logic circuit. The capacitors decouple the power supply and effectively short-out the high frequency spikes. Typical values for the capacitors are $0.002 \, \mu F$ to $0.1 \, \mu F$ depending on the size of the integrated circuit package.

## Pulse reflections

A data highway or bus on a printed circuit when carrying information at very high frequencies can act as a transmission line. If a transmission line is not terminated with its characteristic impedance, reflections can occur from its ends. This may happen if the line is open circuit or terminated by a gate, and a pulse may be reflected back along the line, which will cause any logic connected to it to switch a second time. This behaviour can be observed in high speed counters where reflections on the clock line can cause the counters to switch at twice the expected rate when fast narrow clock pulses are applied. If the pulse width is increased, there may be interaction with the reflections, and then erratic behaviour is observed.

Reflections can be eliminated by terminating the lines with their characteristic impedance by connecting a suitable resistance between the line and earth.

## Cross-talk

Cross-talk can occur when signals running along adjacent tracks or wires are coupled together by mutual capacitance. This can lead to corruption of the data. Forward cross-talk occurs between adjacent tracks where the data signals run in the same direction. It is proportional to the signal edge speed ($dV/dt$) and the length of the coupled tracks. In general, problems usually occur when propagation distances exceed 1 metre.

Back cross-talk occurs when signals travel in opposite directions on adjacent tracks. It can cause serious problems over coupled lengths of a few centimetres.

Cross-talk can only be eliminated by modifying the layout of the circuit. It is reduced by increasing the spacing between tracks on a printed circuit board. It may be necessary to separate cross-talking tracks with a third track at ground potential, or in extreme cases, a ground plane may be needed. If cross-talk occurs between wires, a twisted pair of wires with one at ground potential will afford adequate isolation and in extreme cases screened cable must be used.

$$V = L \frac{di}{dt} \simeq L \frac{\Delta i}{\Delta t}$$
$$= \frac{10^{-4} \times 20 \times 10^{-3}}{10^{-7}} = 20 \, V$$

Typical values 50–300 $\Omega$.

Ground is the same as earth potential.

186

## Summary

Practical logic circuits can be divided into two broad groups. Bipolar logic uses the bipolar transistor, and MOS logic uses the unipolar transistor. Both types of logic can be fabricated as integrated circuits and within each group there is a range of products offering varying speed/power consumption trade-offs. Some bipolar and MOS circuits are incompatible, and additional interfacing circuitry may be required in mixed logic systems.

The number of connections to and from logic gates may have to be restricted so as not to overload the circuits. The manufacturer's data must be adhered to when a logic system is built up.

Logic designs can be implemented within large scale integrated circuits Memories, PLAs and PALs can be programmed as logic systems. Gate arrays can be customized during the last stage of manufacture for specific applications. Libraries of standard cells enable the fast and economical development of logic systems as special purpose integrated circuits.

A digital circuit will invariably have to be interfaced with an analogue system. Data converters may be used for this purpose. In general DACs are simpler and faster than ADCs, as analogue to digital conversion is a serial operation. Recent developments in integrated circuit technology have, however, made parallel ADCs a viable proposition for very fast but low resolution conversions.

Finally, the physical layout of a logic system has a significant bearing on its reliability. The logic designer must decouple power supplies, terminate bus lines and ensure that the circuit layout is satisfactory, with adequate screening, to reduce cross-talk to a minimum.

# Answers to problems

1.1 (i) $100011101_2$, (ii) $0.10111001_2$, (iii) $10101.1101_2$

1.2 (i) $57_{10}$, (ii) $0.171875_{10}$, (iii) $5.671875_{10}$

1.3 (i) 10 bits, (ii) four octal places, (iii) three hex places, (iv) 13 bits

1.4 (i) $266_8$, (ii) $1365_8$, (iii) $57.6_8$

1.5 (i) $B9C7_{16}$, (ii) $127C_{16}$, (iii) $FE.08_{16}$

1.6 (i) $1000111\,0010\,0011_{BCD}$, (ii) $110\,1001\,1001_{BCD}$,
(iii) $100\,0011\,0101\,1001\,1000_{BCD}$

1.7 (i) $100\,1000$, (ii) $111\,1010\,1110$, (iii) $100\,1010$, (iv) $1111\,0100\,1000$

1.8 (i) $1001\,10001$, (ii) $1010\,10001$, (iii) $10010\,01010$, (iv) $11000\,10010$

1.9 Correct data is

| (i) | (ii) |
|-----|------|
| 0001 | 0110 |
| 1111 | 1000 |
| 1100 | 1100 |
| 0000 | 1110 |

1.10 A binary adder may be used, but each decimal stage must be checked to see if either a carry or an invalid BCD code is present (i.e. 1010 to 1111). If so, add 0110 to that stage.

| | |
|---|---|
| Example | $19 + 2 = 21$ |
| In BCD we expect | $1\ 1001 + 10 = 10\,0001$ |
| In binary adder we get | $1\ 1001 + 10 = 1\ 1011$ |
| But least significant group of 4 | 1011 is valid |
| | $\therefore$ add 0110 to least significant BCD group |
| | $1\ 1011 + 0110 = 10\,0001$ |

1.11 Any code where adjacent numbers differ by only 1 bit can be used for position sensing.

1.12 In a weighted BCD code, each bit has a fixed numerical value. This is not the case in a non-weighted BCD code.

2.1 (i) NAND (ii) EX.OR (iii) NAND (iv) OR

2.3

| (i) ABC | F | (ii) ABC | F |
|---------|---|----------|---|
| 000 | 1 | 000 | 1 |
| 001 | 1 | 001 | 0 |
| 010 | 1 | 010 | 0 |
| 011 | 1 | 011 | 0 |
| 100 | 1 | 100 | 0 |
| 101 | 1 | 101 | 0 |
| 110 | 1 | 110 | 0 |
| 111 | 0 | 111 | 0 |

| 2.4 | AB | F |
|---|---|---|
| | 00 | 0 |
| | 01 | 1 |
| | 10 | 1 |
| | 11 | 0 |

$F = A \oplus B$

2.5

| (i) ABC | F | | (ii) ABC | F | | (iii) AB | F |
|---|---|---|---|---|---|---|---|
| 000 | 1 | | 000 | 0 | | 00 | 1 |
| 001 | 1 | | 001 | 0 | | 01 | 1 |
| 010 | 1 | | 010 | 1 | | 10 | 1 |
| 011 | 1 | | 011 | 1 | | 11 | 0 |
| 100 | 1 | | 100 | 1 | | | |
| 101 | 1 | | 101 | 1 | | | |
| 110 | 1 | | 110 | 0 | | | |
| 111 | 0 | | 111 | 1 | | | |

2.8 (ii) $F = \overline{(A + B).\overline{\overline{A}.\overline{C}}}$

2.9 $F = \overline{A}\,\overline{B}C + \overline{A}B\overline{C} + A\overline{B}\,\overline{C} + ABC$

| ABC | F |
|---|---|
| 000 | 0 |
| 001 | 1 |
| 010 | 1 |
| 011 | 0 |
| 100 | 1 |
| 101 | 0 |
| 110 | 0 |
| 111 | 1 |

2.10

| A | B | C | D | Decision |
|---|---|---|---|---|
| 0 | 0 | 0 | 0 | 0 |
| 0 | 0 | 0 | 1 | 0 |
| 0 | 0 | 1 | 0 | 0 |
| 0 | 0 | 1 | 1 | 0 |
| 0 | 1 | 0 | 0 | 0+ |
| 0 | 1 | 0 | 1 | 0+ |
| 0 | 1 | 1 | 0 | 0+ |
| 0 | 1 | 1 | 1 | 1* |
| 1 | 0 | 0 | 0 | 0* |
| 1 | 0 | 0 | 1 | 1+ |
| 1 | 0 | 1 | 0 | 1+ |
| 1 | 0 | 1 | 1 | 1+ |
| 1 | 1 | 0 | 0 | 1 |
| 1 | 1 | 0 | 1 | 1 |
| 1 | 1 | 1 | 0 | 1 |
| 1 | 1 | 1 | 1 | 1 |

* against A

+ against B

2.11

| A | B | C | D | F |
|---|---|---|---|---|
| 0 | 0 | 0 | 0 | 0 |
| 0 | 0 | 0 | 1 | 0 |
| 0 | 0 | 1 | 0 | 1 |
| 0 | 0 | 1 | 1 | 1 |
| 0 | 1 | 0 | 0 | 1 |
| 0 | 1 | 0 | 1 | 0 |
| 0 | 1 | 1 | 0 | 1 |
| 0 | 1 | 1 | 1 | 0 |
| 1 | 0 | 0 | 0 | 1 |
| 1 | 0 | 0 | 1 | 1 |
| 1 | 0 | 1 | 0 | 1 |
| 1 | 0 | 1 | 1 | 0 |
| 1 | 1 | 0 | 0 | 1 |
| 1 | 1 | 0 | 1 | 0 |
| 1 | 1 | 1 | 0 | 1 |
| 1 | 1 | 1 | 1 | 1 |

2.12 Code table

| Base 6 digit | Code A | B | C | D |
|---|---|---|---|---|
| 0 | 0 | 0 | 1 | 1 |
| 1 | 0 | 1 | 0 | 1 |
| 2 | 1 | 0 | 0 | 1 |
| 3 | 0 | 1 | 1 | 0 |
| 4 | 1 | 0 | 1 | 0 |
| 5 | 1 | 1 | 0 | 0 |

Error detector

| A | B | C | D | F |
|---|---|---|---|---|
| 0 | 0 | 0 | 0 | 1 |
| 0 | 0 | 0 | 1 | 1 |
| 0 | 0 | 1 | 0 | 1 |
| 0 | 0 | 1 | 1 | 0 |
| 0 | 1 | 0 | 0 | 1 |
| 0 | 1 | 0 | 1 | 0 |
| 0 | 1 | 1 | 0 | 0 |
| 0 | 1 | 1 | 1 | 1 |
| 1 | 0 | 0 | 0 | 1 |
| 1 | 0 | 0 | 1 | 0 |
| 1 | 0 | 1 | 0 | 0 |
| 1 | 0 | 1 | 1 | 1 |
| 1 | 1 | 0 | 0 | 0 |
| 1 | 1 | 0 | 1 | 1 |
| 1 | 1 | 1 | 0 | 1 |
| 1 | 1 | 1 | 1 | 1 |

3.1  $F_1 = \overline{A}\,\overline{B}\,\overline{C} + \overline{A}\,\overline{B}C + \overline{A}B\overline{C} + AB\overline{C} + ABC$
      $F_2 = \overline{A}BC + A\overline{B}\,\overline{C} + A\overline{B}C + AB\overline{C} + ABC$

3.2  (i) $F_1 = A\overline{C} + \overline{B}C$
        $F_2 = \overline{A}B\overline{C} + A\overline{C}D + ABC + \overline{A}CD$
        $F_3 = \overline{A}.\overline{B}$
     (ii) $F_1 = (\overline{A} + C).(\overline{B} + \overline{C})$
        $F_2 = (A + B + C).(\overline{A} + C + D)(A + \overline{C} + D)(\overline{A} + B + \overline{C})$
        $F_3 = \overline{A}.\overline{B}$

3.3  $F_1 = \overline{\overline{\overline{AB}.\overline{BC}.\overline{ACD}.\overline{AB}\overline{D}}}$
      $F_2 = \overline{B.\overline{\overline{A}\,\overline{C}}}$

3.4  $F_1 = \overline{\overline{(\overline{A} + \overline{C}) + \overline{(B + \overline{C})}}}$
      $F_2 = \overline{\overline{(\overline{A} + \overline{B})} + \overline{(\overline{A} + C)} + \overline{(B + C)} + \overline{A + B + D}}$

3.5  1st canonical form: $F = \overline{A}B + A\overline{D}$
      2nd canonical form: $F = (A + B).(\overline{A} + \overline{D})$

3.6  $W = \overline{A}.\overline{BD}.\overline{BC}$
      $X = \overline{\overline{\overline{BD}}.\overline{\overline{BC}}.\overline{BCD}}$
      $Y = \overline{\overline{CD}.\overline{\overline{C}D}}$
      $Z = D$

3.7  $W = A$
      $X = A\overline{B} + \overline{A}B$
      $Y = B\overline{C} + \overline{B}C$
      $Z = C + \overline{C}D$
      Three EX.OR gates required.

3.8  $F = \overline{\overline{\overline{AD}.\overline{BD}.\overline{CD}.\overline{A}BC.\overline{ABC}.AB\overline{C}}}$

3.9  $F = \overline{\overline{(A + B + C)} + \overline{(A + \overline{B} + D)} + \overline{(\overline{B} + C + D)} + \overline{(\overline{A} + B + \overline{C} + \overline{D})}}$

3.10 Hazard-free circuit is: $F = AB + \overline{A}CD + \overline{A}\overline{B}C + BCD + \overline{A}\overline{B}D$

3.11 $F = BC + \overline{A}CD + \overline{A}C\overline{E} + \overline{B}\,\overline{C}\,\overline{D}E$

3.12 This problem does not have a unique solution. Answer depends on the cell specification.

3.13 $F = \overline{A}B + AD + CD + B\overline{C}\,\overline{D}$

3.14 See answer 3.11.

4.1  Stable states are ①, ②, ③, ④, ⑤.

4.2

| $b_1$ | $b_0$ | S | R | Q |
|---|---|---|---|---|
| 0 | 0 | 0 | 0 | 0 |
| 0 | 1 | 1 | 0 | 1 |
| 1 | 0 | 0 | 1 | 0 |
| 1 | 1 | 0 | 0 | 0 |

4.3

| $b_1 = J$ | $b_0 = K$ | Q |
|---|---|---|
| 0 | 0 | 0 |
| 0 | 1 | 0 |
| 1 | 0 | 1 |
| 1 | 1 | 0 |

Timing problems eliminated by delaying counter clock.

4.4 Nine gates.

4.5 If $AB = 11$ and internal state $F_1'F_2'$ is either 01 or 11, then $F_1F_2$ oscillates between 01 and 11.

4.6 Use a flow table to show that there is no transition from $F = 1$ to $F = 0$.

4.7

| $Q_2$ | $Q_1$ | $Q_0$ |
|---|---|---|
| 0 | 0 | 0 |
| 0 | 0 | 1 |
| 0 | 1 | 0 |
| 1 | 0 | 1 |
| 1 | 1 | 0 |
| 0 | 0 | 1 |

4.8 $S_1 = A$, $R_1 = Q_1Q_2B + C$, $S_2 = A$, $R_2 = C$.

Protection circuit. AND each push button with the inverse of all the others before inputting to controller, e.g. $A_c = A\overline{B}\overline{C}$.

4.9

| $Q_2$ | $Q_1$ | $Q_0$ |
|---|---|---|
| 0 | 0 | 0 |
| 0 | 0 | 1 |
| 0 | 1 | 0 |
| 1 | 1 | 1 |

4.10

| $Q_2$ | $Q_1$ | $Q_0$ |
|---|---|---|
| 0 | 0 | 0 |
| 0 | 0 | 1 |
| 0 | 1 | 0 |
| 1 | 1 | 1 |
| 1 | 0 | 0 |
| 0 | 0 | 1 |

4.11

| $Q_2$ | $Q_1$ | $Q_0$ |
|---|---|---|
| 0 | 0 | 1 |
| 0 | 1 | 0 |
| 1 | 0 | 1 |
| 0 | 1 | 1 |
| 1 | 1 | 1 |
| 1 | 1 | 0 |
| 1 | 0 | 0 |
| 0 | 0 | 1 |

4.12

| $Q_3$ | $Q_2$ | $Q_1$ | $Q_0$ |
|---|---|---|---|
| 0 | 0 | 0 | 0 |
| 0 | 0 | 0 | 1 |
| 0 | 0 | 1 | 1 |
| 0 | 1 | 1 | 1 |
| 1 | 1 | 1 | 1 |

5.1 $T_0 = I$; $T_1 = Q_0\overline{Q}_3$; $T_2 = Q_1$; and $T_3 = Q_2 + Q_0Q_3$. (No output logic required.)

5.2 Output logic is $Z_3 = Q_2$, $Z_2 = Q_2\overline{Q}_0 + Q_2\overline{Q}_1 + \overline{Q}_2Q_1Q_0$, $Z_1 = Q_2 \oplus Q_1$ and $Z_0 = \overline{Q}_2Q_1 + \overline{Q}_2Q_0 + Q_2\overline{Q}_1Q_0$.
    Shift register version: use 4-bit shift register, inverted feedback and starting state 0000.

5.3 Driving logic for JKFFs $J_1 = Q_0$, $K_1 = \overline{Q}_0$, $J_0 = 1$, $K_0 = Q_1$
                    for TFFs $T_1 = Q_1 \oplus Q_0$, $T_0 = \overline{Q}_0 + Q_1$
                    for DFFs $D_1 = Q_0$, $D_0 = Q_0 + \overline{Q}_1$

5.4 Either design a four-state 3-bit counter to output $RAG = 100,110,001$, $010,100$, etc., or use a 2-bit pure binary counter and output logic: $Z_R = \overline{Q}_1$, $Z_A = Q_0$, $Z_G = Q_1\overline{Q}_0$.
    Use the control waveform to drive the counter clock.

5.5 $D_2 = IQ_2 + Q_2Q_0 + Q_2Q_1 + IQ_1Q_0$
    $D_1 = I\overline{Q}_1Q_0 + Q_2Q_1Q_0 + \overline{I}\overline{Q}_1Q_0 + IQ_1\overline{Q}_0 + \overline{I}Q_2\overline{Q}_1\overline{Q}_0$
    $D_0 = Q_1\overline{Q}_0 + Q_2\overline{Q}_0 + I\overline{Q}_0 + IQ_2Q_1$

5.6 For faulty JKFF

| J | K | $Q_t \to Q_{t+1}$ |
|---|---|---|
| 0 | d | $0 \to 0$ |
| 1 | d | $0 \to 1$ |
| 0 | d | $1 \to 1$ |
| 1 | d | $1 \to 0$ |

    Ignore K and relabel J as T and sell as synchronous TFFs.

5.7 Driving logic: $J_1 = IQ_0$, $K_1 = 1$, $J_0 = I$, $K_0 = \overline{I} + Q_1$
    System generates code sequence 00, 01, 11, 00, etc., providing input $I = 1$.
    If $I = 0$ then reset to 00.

5.8 $D_1 = I_1$
    $D_0 = I_0$
    $Z = I_1\overline{Q}_0 + I_0\overline{Q}_1\overline{Q}_0 + I_1I_0\overline{Q}_0$

5.9

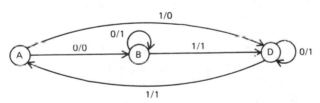

5.10 No unique solution. Obtain state transition diagram, state-minimize to four internal states, obtain present/next state table and then design logic.

5.11 No unique solution. Proceed as in Problem 5.8. System minimizes to four internal states. Use an additional input push button (enter) to synchronize the input and clock the flip-flops, if synchronous devices are used.

5.12 No unique solution. This system minimizes to seven internal states.

| 6.1 Multiplexer inputs | $D_0$ | $D_1$ | $D_2$ | $D_3$ | $D_4$ | $D_5$ | $D_6$ | $D_7$ |
|---|---|---|---|---|---|---|---|---|
| (i) | 0 | 0 | 1 | 0 | 1 | 1 | 0 | 1 |
| (ii) | 1 | 1 | 0 | 1 | 1 | 0 | 0 | 1 |
| (iii) | 0 | 0 | 0 | 0 | 0 | 1 | 0 | 0 |
| 6.2 (i) | 0 | 1 | 0 | $\overline{A}$ | A | $\overline{A}$ | 1 | 0 |
| (ii) | A | A | A | A | 0 | 0 | A | A |

6.3

6.4 No unique solution. One simple method is to encode votes into 2 bits – for = 10, against = 01 and abstain = 00 – and form an 8-bit word for the four voters. Address a $2^8$ (256)-bit memory with the 8-bit word. Set each location in memory to 1 if its address word contains more odd bits ($b_7$, $b_5$, $b_3$, $b_1$) set to 1 than even ($b_6$, $b_4$, $b_2$, $b_0$).

6.5 Obtain present/next state table. Counter increments if $I = 1$ and halts if $I = 0$. Memory must have six address inputs (I and five output bits fed back). A 64 × 5-bit memory is required. Store present/next state table in memory. Feedback must be via DFFs in order to synchronize the system. Reset the system via clears on DFFs.

6.6 4 × 6 × 3 PLA.

6.7 No unique solution. $F_1$ and $F_2$ must be reduced to no more than three implicants between them. BC is common between $F_1$ and $F_2$.

6.8 No unique solution.

6.9

| Fault | Test ABC | Result |
|---|---|---|
| a | 010 | $F = 0$, $F_a = 1$ |
| b | 100 | $F = 0$, $F_b = 1$ |
| c | undetectable | |
| d | 100 | $F = 0$, $F_d = 1$ |
| e | 000 | $F = 0$, $F_e = 1$ |
| f | 010 | $F = 0$, $F_f = 1$ |
| g | or | $F = 0$, $F_g = 1$ |
| h | 100 | $F = 0$, $F_h = 1$ |

6.10

| Fault | Test ABC | Result |
|---|---|---|
| a | 110 | $F = 1$, $F_a = 0$ |
| b | 110 | $F = 1$, $F_b = 0$ |
| c | undetectable | |
| d | undetectable | |
| e | 110 | $F = 1$, $F_e = 0$ |
| f | undetectable | |

$$\begin{array}{lll}
\text{g} & \left\{\begin{array}{l} 001 \\ \text{or } 011 \end{array}\right. & F = 1, \quad F_g = 0 \\[12pt]
\text{h} & \left\{\begin{array}{l} 101 \\ \text{or } 110 \\ 111 \\ 101 \\ 100 \end{array}\right. & F = 1, \quad F_h = 0
\end{array}$$

Note. Faults a, b, and e are indistinguishable.

7.1  $S = \overline{\overline{A}\,\overline{B}C_{in}} \; \overline{\overline{A}B\overline{C_{in}}} \; \overline{A\overline{B}\,\overline{C_{in}}} \; \overline{ABC_{in}}$

   $C(out) = \overline{\overline{A}\,\overline{B}} \; \overline{\overline{A}\,\overline{C_{in}}} \; \overline{\overline{B}\,\overline{C_{in}}}$

7.2  $S = \overline{\overline{(A + B + C_{in})} + \overline{(A + \overline{B} + \overline{C_{in}})} + \overline{(\overline{A} + \overline{B} + C_{in})} + \overline{(\overline{A} + B + \overline{C_{in}})}}$

   $C(out) = \overline{\overline{(A + C_{in})} + \overline{(A + B)} + \overline{(B + C_{in})}}$

7.3  $C_2(out) = A_2 B_2 + A_2 \overline{B}_2 A_1 B_1 + \overline{A}_2 B_2 A_1 B_1 + A_2 \overline{B}_2 A_1 B_1 A_0 B_0$
   $+ \; \overline{A}_2 B_2 A_1 B_1 A_0 B_0 + A_2 \overline{B}_2 \overline{A}_1 B_1 A_0 B_0 + \overline{A}_2 B_2 \overline{A}_1 B_1 A_0 B_0$

7.4  (i) $1010111_{TC}$, (ii) $1010011_{TC}$, (iii) $1010001_{TC}$, (iv) $100011100_{TC}$

7.5  $D_0 = A_0 \overline{B}_0 + \overline{A}_0 B_0$
   $D_1 = \overline{A}_1 \overline{A}_0 \overline{B}_1 B_0 + A_1 \overline{A}_0 B_1 B_0 + A_1 \overline{B}_1 \overline{B}_0 + A_1 A_0 \overline{B}_1 + \overline{A}_1 A_0 B_1 + \overline{A}_1 B_1 \overline{B}_0$
   $D_2 = \overline{A}_1 \overline{A}_0 B_0 + \overline{A}_0 B_1 B_0 + \overline{A}_1 B_0$

7.6  A $16 \times 3$-bit memory is required – the address input is $A_1 A_0 B_1 B_0$ and the outputs are $D_0$, $D_1$ and $D_2$. The contents of the memory are the truth tables for functions $D_0$, $D_1$ and $D_2$ of Problem 7.5.

7.7  Initially a $256 \times 5$-bit memory is required. The address is formed from the two BCD inputs (8 bits) and the BCD sum + carry-out (5 bits) stored in the memory. If an external memory coder is used, the memory can be reduced because not all input addresses are valid BCD codes. Further, if use is made of the commutative properties of addition and only one entry is allowed for A + B and B + A, the memory can be reduced to $64 \times 5$-bits resolution of which 55 addresses need to be used.

7.8  No unique solution.

7.9  If the input is $A_3 A_2 A_1 A_0$, the output functions are

$$S_0 = A_0$$
$$S_1 = 0$$
$$S_2 = A_1 \overline{A}_0$$
$$S_3 = A_2 \overline{A}_1 A_0 + \overline{A}_2 A_1 A_0$$
$$S_4 = A_2 \overline{A}_1 \overline{A}_0 + \overline{A}_3 A_2 A_0 + A_3 \overline{A}_2 A_0$$
$$S_5 = A_3 A_2 A_0 + \overline{A}_3 A_2 A_1 + A_3 \overline{A}_2 A_1$$
$$S_6 = A_3 \overline{A}_2 + A_3 A_1$$
$$S_7 = A_3 A_2$$

7.10 Valid subtraction always produces an overflow signal (carry-out) of $1_2$ from the adder. If carry-out is $0_2$ negative overflow has occurred, i.e. negative overflow is inverse of carry-out.

# Index

Page numbers in **bold** refer to figures and page numbers in *italic* refer to tables.